Two Souls Indivisible

Books by James S. Hirsch

HURRICANE
The Miraculous Journey of Rubin Carter

RIOT AND REMEMBRANCE
The Tulsa Race War and Its Legacy

TWO SOULS INDIVISIBLE
The Friendship That Saved Two POWs in Vietnam

TWO SOULS
INDIVISIBLE

The Friendship That
Saved Two POWs in Vietnam

James S. Hirsch

HOUGHTON MIFFLIN COMPANY

BOSTON · NEW YORK 2004

For information about permission to reproduce selections from
this book, write to Permissions, Houghton Mifflin Company,
215 Park Avenue South, New York, New York 10003.

Visit our Web site: www.houghtonmifflinbooks.com.

ISBN-13 978-0-618-27348-5
ISBN-10 0-618-27348-4

Library of Congress Cataloging-in-Publication Data

Hirsch, James S.
Two souls indivisible : the friendship that saved
two POWs in Vietnam / James S. Hirsch.
p.   cm.
ISBN 0-618-27348-4
1. Vietnamese Conflict, 1961–1975 — Prisoners and
prisoners, North Vietnamese. 2. Cherry, Fred V.
3. Halyburton, Porter. 4. Prisoners of war — United
States. 5. Prisoners of war — Vietnam. I. Title.
DS559.4.H57 2004
959.704'37'092273 — dc22   2003067595

PRINTED IN THE UNITED STATES OF AMERICA

Book design by Robert Overholtzer

1 2 3 4 5 6 7 8 9 10 QUM

To Amanda and Garrett,

Pearls in the Constellation

CONTENTS

TWO SOULS
INDIVISIBLE

1

"Better Place, Worse Place"

Better place, worse place."

Eagle slammed the notebook closed and gave the young American prisoner of war an ultimatum: talk to him and be taken to a camp where he could be with his buddies or refuse to cooperate and be taken to a place where he would suffer. Captured only a few days earlier, U.S. Navy Lieutenant (junior grade) Porter Halyburton didn't know the consequences if he continued to withhold military information. He was already locked inside North Vietnam's notorious Hoa Lo Prison, dubbed "the Hanoi Hilton" by the Americans, a forbidding trapezoidal structure with thick outer walls topped by barbed wire and jagged glass. Years of urine, blood, and vomit permeated the rotting crevices. The food included chicken feet and bread so moldy that it had begun to ferment. Even the prison's name suggested its hellishness — Hoa Lo (pronounced "wa-*low*") means "fiery furnace" in Vietnamese.

Whatever was "worse" would certainly be terrible, Halyburton thought, but still not as abhorrent as assisting the enemy.

At twenty-four, Halyburton was one of the younger American POWs in Vietnam. His six-foot frame, short brown hair, and wholesome good looks fit the prototype of the dashing "fighter jock," whose love of danger and combat had been immortalized in film and literature. But Halyburton was also introspective and artistic, the product of a small college town that had nurtured his intellectual and creative pursuits. He wrote poems, carved wooden statues, and read widely on history and culture. He was also a family man, having married his college sweetheart. The couple's baby daughter was born four weeks before he left for Vietnam.

He was lucky to be alive. On October 17, 1965, his F-4 Phantom jet was shot down forty miles northeast of Hanoi, killing the pilot in a fiery explosion. Halyburton, the "backseat" navigator, ejected without injury. Among many combat aviators, it was an article of faith that they would rather die instantly in a crash than be caught by the enemy. Halyburton believed otherwise, but he soon realized that the price of survival would be high.

Immediately after his capture he was sent to Hoa Lo, where his cell, seven feet by six, had a boarded window, a single dim light bulb, and a concrete bed with leg irons. Cockroaches darted through the cells, and rats, some over a foot long, prowled the premises, lending evidence to a postwar POW study that noted, "After sundown, rats and mice literally took over North Vietnam." Scribbled across the faded whitewashed walls were Vietnamese letters, but so too was something more comforting — the name of an American, Ron Storz. Halyburton wasn't isolated or completely deprived; he could whisper to Americans in adjoining cells and was allowed to shower. Interrogations became a part of daily life: he was questioned by Colonel Nam, a gray-haired Vietnamese commander called Eagle for his authoritarian manner. Using passable English, he offered Halyburton the carrot or the stick. It was his choice.

"Better place, worse place," Eagle intoned repeatedly.

Halyburton only disclosed the information prescribed by the

Code of Conduct for captured American servicemen: "Porter Halyburton," he said. "Lieutenant j.g., 617514, 16 January 1941."

Further "quizzes," as they were called, produced the same response, so after two weeks a guard went to Halyburton's cell one night, blindfolded and handcuffed him, and walked him to a truck, which rumbled a couple of miles to the outskirts of Hanoi. He was left at the Cu Loc Prison, believed to be a former French film studio where the grounds were still littered with old film cans, ducks and chickens roamed, and mosquitoes buzzed. A large putrid swimming pool lay thick with water, dirt, garbage, and fish that the Vietnamese guards raised for food. When Halyburton was pushed into his pitch-black cell, he pressed his hands against the walls to discover its dimensions. The room, though relatively large — each wall was fifteen feet long — was indeed worse than his previous cell. There was no bed, no light, its window was bricked up, and it smelled of wet concrete. But at least Halyburton could still use a tap code to communicate with the POWs in adjacent cells. He was not alone.

The harassment, however, continued. In the quiz room, Halyburton sat on a stool that forced him to look up at his new interrogator, a surly, jug-eared official nicknamed Rabbit, who called the American an "air pirate" and a "war criminal." He made the same threat — "better place, worse place" — if Halyburton did not reveal the names of his ship, squadron, and plane, but the prisoner didn't give in. The threat was fulfilled: days later, he was moved across the compound to a remote storage room in the back of an auditorium. Once again feeling his way in the darkness, he discovered that this space was only five feet by eight. What's more, it was isolated, preventing any communication with other Americans. That scared him. Except for interrogations, the only time he left the cell was to empty his waste bucket, and there was no more bathing. The questioning had become more abusive; Halyburton was repeatedly harangued ("Bad attitude! Bad attitude!") and slapped across the head.

He sought comfort through prayer. He did not ask for freedom, for food, or for any material comforts. He asked for strength to survive, for companionship, and for the safety of his family. He found inspiration, literally, from above.

One morning he noticed a beam of sunlight filtering through the shutters in his cell and arcing across his cement wall. The next morning he saw the light strike the same place. So he tore a piece of coarse brown toilet paper into the shape of a cross and used rice to stick it on the cement. The following morning the light slowly passed over the cross — a radiant signal from God, Halyburton thought. He gratefully whispered the Lord's Prayer.

But the solace didn't last. Halyburton continued to refuse to provide military information and was again taken to a "worse place," this time to a nearby shed. It had two rooms, but he was confined to one that was again five feet by eight. The place had once stored coal and was later designated by the Americans as the "outhouse" or "shithouse." A few holes in the ceiling and space beneath the door supplied scant ventilation, and coal dust covered the floor. Through cracks in the wall, Halyburton could see other Americans walking together in the compound, and he didn't understand why he had been singled out for isolation and mistreatment. Had the other POWs cooperated with the enemy to receive better treatment? In captivity for a month, he had lost twenty-five pounds and had developed dysentery. It was now late November and cold, and his mosquito net provided flimsy refuge from the insects' nightly assaults. The interrogations also continued: Halyburton was questioned about his life as well as the war.

"Where do you live?"

"What are your parents' names?"

"Do you have a family?"

By now, the Vietnamese had discovered on Halyburton's flight vest the names of his squadron and ship, and they knew that he was married and from North Carolina, which he assumed they had learned from U.S. newspapers. That information, in enemy hands, felt like one more violation, and Halyburton feared he was breaking down mentally as well as physically.

But he hadn't broken, and he still refused to answer questions beyond his name, rank, serial number, and date of birth. Rabbit presented the familiar choice: "Better place, worse place."

Halyburton didn't respond and was taken back to his filthy cell.

He slumped down in despair. He doubted the Vietnamese would purposely kill him. Dead, he was useless to them; alive, he could still, in theory, provide military information or propaganda statements. But Halyburton knew he could perish from abuse or neglect, and it occurred to him that his isolation could doom him to an ignominious end. He could die in his cell, quietly, with the geckos, rats, and mosquitoes whose musty space he had shared. His death would be one more inconvenience for his Vietnamese guard, who twice a day received rations for the prisoner but waited at least an hour before sliding the food in, allowing ants to infest the rice and cool air to congeal the pig fat in the watery soup. His death would be his final deliverance, but beyond the enemy, who would even know?

The lock turned and the wooden door swung open, allowing the guard and a commander to enter. It was November 28, nighttime, forty-two days after Halyburton's plane had been shot down. He knew that a visit at this hour meant he would be moved to another cell — a worse place — but he wasn't sure how much more he could endure. He used his blanket to roll up his mosquito net, some clothes, a tin cup, and his toothbrush, and he followed the guard and interrogator through the compound. The air was cool and refreshing, and the soft grass massaged his bare feet.

Something was alive, he thought, something that wasn't caked with dirt. They walked about thirty yards, turned left, and approached a one-story building known as "the Office," whose five rooms had been converted to prison cells. It was, in fact, the same building he had initially been taken to. They went up two concrete steps and reached the door to cell number one. Halyburton's mind raced with thoughts about the misery that awaited him. What could be worse than a dark, claustrophobic room with coal dust, rats, and lizards?

The door opened, and Halyburton walked inside. A faint bulb emitted just enough light for him to see a man sitting on a teak board that served as a bed. He was thin, unwashed, unshaven, and injured, his left foot wrapped in a cast and his left arm hanging in a sling. He was black.

"You must take care of Cherry," the guard said.

The door was slammed shut. After a long pause, the newcomer stepped forward.

"I'm Porter Halyburton. I'm a Navy j.g. F-4. Backseater."

"Major Fred Cherry," the black officer said. "Air Force. F-105 Thunderchief."

Halyburton soon realized that his new torment had nothing to do with grimy cells, unpalatable food, or sadistic guards. His new punishment — the "worse place" — was to care for a black man.

The Vietnam War was the longest in U.S. history and, with more than 58,000 Americans killed, the third deadliest. It was also a wrecking ball through American society, igniting passionate protests in town squares and campuses, radicalizing a youth movement, tormenting political leaders, and stymieing a great military that could not subdue a peasant nation. It spawned cynicism toward public institutions, disdain for veterans, and doubt about America's role in the world. By the time the war ended in January of 1973, most Americans had concluded that the effort had been ill defined and poorly executed, and the country would spend the rest of the century debating "the lessons of Vietnam."

But on one matter there was no debate — the POWs. When the Democratic Republic of Vietnam released 591 U.S. prisoners at war's end, their return represented a singular accomplishment in a conflict without defining victories or tangible gains. The POWs' sacrifice, perseverance, and patriotism were celebrated by countrymen whose faith in the armed services and in America itself had been shaken. The returning prisoners were feted at the White House, saluted at homecoming parades, and acclaimed as heroes. California's governor Ronald Reagan said: "You gave America back its soul — God bless a country that can produce men like you."

For all the attention they received, the number of POWs in Vietnam was actually quite small compared to those from the century's other major wars (130,201 in World War II, for example, and 7,140 in the Korean War). Yet the fate of the Vietnam prisoners was a national melodrama, driven in part by the POWs' wives, who orchestrated a savvy publicity campaign that pressured the country to place their husbands' return at the center of any peace accord. The POW bracelet, launched by a private organization, was another brilliant publicity gambit that allowed Americans to view the captives as individuals and support them without endorsing the war itself.

Of course, some of the captured Americans did not return. At least eighty-four died in Southeast Asian prisons and jungle camps, usually from torture, untreated wounds, or execution. But the survival rate was high, given the abject living conditions and the sheer length of their confinement. Unlike common criminals in civilian prisons, the POWs were not serving a defined sentence. Their confinement was unknown and indefinite. Until Vietnam,

7

no U.S. military prisoner had been held in captivity for more than four years, but the Vietnam War saw more than three hundred Americans incarcerated for five or more years; two men were held for nine years. Their experience had no precedent in American history.

The prisoners in Hanoi had a very different profile from those of the grunts fighting in South Vietnam. They were professional soldiers and tended to be older college graduates whose maturity and experience sustained them through the lowest moments of their ordeal. These officers found unity and strength by developing an elaborate military command structure, a secret communications network, and a rigorous code of conduct, and many returned with extraordinary tales of survival, overcoming years of abuse and privation while finding value in their own suffering.

But in the many personal narratives of courage and defiance, the story of Porter Halyburton and Fred Cherry stands apart. They were locked in the same cell because the Vietnamese believed their racial differences would torment them — a not entirely naïve assumption. While the two officers were separated by age, rank, and military service, each man's race had produced a dramatically different life experience. Cherry, descended from a Virginia slave, was a pioneer in the integration of the armed services; though sustaining many racist slights along the way, he became one of the Air Force's best combat pilots. Halyburton, whose forefathers fought for the Confederacy, was raised in the segregated South, where blacks were poor, deferential, and inferior; his was not the virulent racism of the demagogue but the more insidious bigotry of condescension and paternalism.

Each man, ultimately, carved a distinctive legacy in Vietnam during a confinement of seven and a half years. Cherry was renowned for his resolve against the Vietnamese, who showed no mercy in trying to convince him that he should repudiate "the American imperialists" and support the colored people of Asia. Cherry suffered as much physical pain as any prisoner who survived, yet he appears to be the only tortured POW who never made concessions to the enemy. Halyburton was respected as a creative scholar, who invented such games as invisible bridge — played without cards — and whose imagination allowed him to find a meaningful life in the bleakest of settings.

Halyburton and Cherry returned home to very different circumstances, which mirrored the range of experience for all the POWs on their repatriation. Halyburton's wife, Marty, was initially told that he had been killed in action, and a memorial service was held to honor his memory. Sixteen months later, learning he was alive, she remained loyal to him, speaking out on his behalf and becoming stronger and more independent from the adversity. But Cherry's marriage, already on shaky ground when he was captured, did not survive. His wife quickly turned on him, spent his money, and splintered the family. Both the Halyburton and Cherry families learned, through years of estrangement, fear, and hope, that the inmates in Hanoi were not the only prisoners. "We were all POWs," said Cherry's son, Fred Jr.

Halyburton and Cherry were in the same cell for less than eight months. They were grateful to have a roommate, though each was initially wary of the other. Cherry thought Halyburton was a French spy, while Halyburton doubted that a black could be a pilot. But they overcame their misgivings and preconceptions and found common ground in this uncommon environment — a friendship in extremis that inspired many of their fellow prisoners. As Giles Norrington, a Navy pilot shot down in 1968, recalled, "By the time I arrived, Porter and Fred had already achieved legendary status . . . The respect, mutual support, and affection that had developed between them were the stuff of sagas. Their stories, both as individuals and as a team, were a great source of inspiration."

Many of the POWs had to cross racial, cultural, or social boundaries to exist in such close confines. But Halyburton and Cherry did more than coexist — they rescued each other. Each man credits the other with saving his life. One needed to be saved physically; the other, emotionally. In doing so, they forged a brotherhood that no enemy could shatter.

2

One More Round

Fred Vann Cherry was a five-foot-seven flying ace, built like a whip, whose calm demeanor and steady nerves were required in and out of the cockpit. Entering the Air Force in 1951, he was a pioneer in the military's integration, a black officer who performed with distinction in the Korean War, manned critical posts at the height of the Cold War, and now, in 1965, was leading bombing raids in Southeast Asia. Yet he was still an anomaly — the Air Force had only twenty-one hundred black officers, 1.6 percent of the total — and over the years he had faced many racial snubs, some overt, some subtle. His response was always the same: to turn the other way, to ignore them, to never jeopardize his standing in the Air Force. In short, to keep quiet.

He had been in Japan since 1961, serving with the 35th Tactical Fighter Squadron, and had spent three years rotating into South Korea, where he had week-long assignments sitting on "nuclear alert." If ordered, he could be airborne in less than four minutes, his job to fly over enemy territory and drop a nuclear weapon. Despite the high stakes, the assignment was insufferably dull, forcing Cherry to sit in a room wearing his flight suit for days at a time, playing Ping-Pong or poker, watching movies, and waiting. When his week in Korea was over, he would return to an air base in Japan, where he continued training until he was sent back on alert.

Cherry was supposed to return to the United States in 1963, but he asked for an extension because he wanted to fly a new jet fighter, the F-105 Thunderchief, the fastest tactical plane in the Air Force and able to fly nearly 1,800 miles without refueling. The 8th Tactical Fighter Wing, which consisted of three squadrons, including Cherry's, was grateful that Fred remained in Japan. He became an expert on the Thunderchief's weapons; he was selected to write the plane's guidebook for the wing, and he also wrote a favorably reviewed article about the F-105 for a military publication, *Pacaf Flyer.*

Staying in Japan suited Cherry's wife and their four children. While they often faced bigotry in America — they moved overseas before the landmark civil rights legislation of 1964 — they now lived in relative harmony on American air bases. The family had a beloved *mama-san,* who cooked for them, washed and ironed their clothes, and bathed the children. Cherry rode his kids on a motor scooter, taking them to baseball practice and judo, karate, and swimming lessons. (The last was particularly important to Fred, who couldn't swim a stroke, even though he often flew over water.) The military's schools had excellent American teachers, and on Christmas Eve a helicopter would land in the American compound and drop off Santa Claus, bearing gifts for all the children.

There were other benefits in Japan, whose citizens appreciated America's role in rebuilding the country after it had been conquered in World War II. Other countries, they believed, would have treated them harshly, whereas U.S. servicemen now protected them. Of course, the servicemen's money also made them popular.

The combat pilots themselves had a cultlike following, their aerial bravado inspiring respect, even awe. Who else flew high-powered, multimillion-dollar jets through the dark skies, only to encounter Soviet MiGs or antiaircraft fire or surface-to-air missiles, before dropping their own bombs in the name of freedom and democracy. American aviators were considered "the tip of the spear" for the entire fighting force, and when they swaggered through the doors of a club or restaurant, crowds parted and eyes widened. Their uniforms alone projected a kind of macho authority: olive jumpsuits zipped up the front gripped every muscle and were creased at the crotch from parachute straps, while zippered pockets lined the legs, arms, and chests. Unshaven, their faces and hands streaked with grime and sweat from their latest mission, the airmen spoke loudly, caroused freely, and reveled in their own glory. They had a saying, most often uttered after several rounds of drinks: "A good fighter pilot can outfight, outfly, and outfuck anyone else in the world."

Cherry enjoyed this sybaritic life. In the Korean War, he discovered that prostitution was legal there, though some "cathouses" didn't allow blacks. By 1965, prostitution had been nominally outlawed in Korea, but brothels continued to flourish. Japan had plenty of attractions as well. Bachelors in the military lived in rented mansions that accommodated raucous, glass-shattering parties, complete with drinking contests, fistfights, and attractive women.

The womanizing was part of a military subculture, particularly in Asia, where mistresses were common and infidelity the norm; the men who risked their lives were considered entitled. As Ellsworth Bunker, a U.S. ambassador in Saigon during the Vietnam War, observed, "There's a lot of plain and fancy screwing going on around here, but I suppose it's all in the interest of the war effort."

The military wives, of course, believed otherwise, and Cherry's philandering contributed to the tensions in his own marriage. But for a man whose race made him an outsider, his embrace of the military's bacchanalian customs contributed to his acceptance among his peers.

Cherry's greatest passion was piloting jet fighters, and in this sense his decision to stay in Japan was vindicated. In 1964 the Kennedy administration, seeking to thwart the Communist insurgents in Indochina, increased its military personnel in South Vietnam from 10,000 to 23,000. It also called for air raids into North Vietnam and Laos, inching America into a full-scale but undeclared war.

Cherry had not seen combat since the Korean War, where he flew fifty-two sorties, received two air medals, and was part of the 58th Fighter Bomber Group, which received a Distinguished Unit Citation for "extraordinary heroism." He passed the succeeding years training, instructing, and simulating attacks, earning top marks for gunnery and bombing, and receiving promotions and praise. In the middle 1950s, at Malmstrom Air Force Base in Great Falls, Montana, the wing commander was a broad-shouldered colonel named Murray Bywater. When he flew himself to air bases across country, he had to be accompanied by a second aircraft, its pilot responsible for the flight plan. Bywater chose Fred Cherry, surprising many that he would pick a black man for such a highly visible job. But as he recalled years later, Cherry "was the best pilot in the wing."

By 1964 Cherry was a flight leader, who exercised significant control on a mission. A single flight consists of four aircraft in staggered formation, with two leads followed by two wingmen. This synchronization provides maximum support and protection for the entire flight, but it also places the power in the hands of the lead pilots: where they go, their wingmen go.

Cherry's race increased his pressures to perform. Before 1948, the military had segregated blacks for many reasons; not least was the belief that they were unfit to lead whites into battle. In 1964 the dearth of black officers ensured that it rarely happened. Cherry was an exception, and he gladly defied the racist stereotypes of black commanders. Not only did he lead whites into battle, whites pleaded to be in his flight, just as white students asked to be in his gunnery classes. Cherry's nickname, Chief, connoted his authority and respect. He also dazzled his commanders — one said he moved through the air "like an eel." Major Bobby J. Mead, in an evaluation, wrote on March 6, 1964: "I consider Captain Cherry one of the most effective officers of his rank that I have worked with during my entire Air Force career." As Ed Kenny, one of his early gunnery instructors, said, "Fred always had that little man in him that kept wanting him to do better."

For Cherry's part, social statements were incidental to his ambition. What motivated him was the excitement of airborne combat, in which do-or-die engagements were the ultimate test of skill, daring, and courage. Like all great fighter pilots, he never had any qualms about his work. He believed that if a pilot couldn't pull the trigger, he should fly cargo planes. Cherry otherwise had few hobbies, pastimes, or interests. Flying combat missions was what he did best, and Vietnam gave him one more chance.

While the U.S. Navy could send jets from carriers off the coast of Vietnam, the Air Force could not do the same from distant Japan. It needed cooperation from Thailand, where in 1964 the Americans turned primitive air fields in Korat and Takhli into crude bases. Air Force personnel built wooden hooches on stilts to avoid the cobras, waded through six-inch puddles that formed in minutes from fierce downpours, brought in air-conditioned trailer homes for senior officers, and cut through the thick vegetation that covered the runway lights. The work was difficult and sometimes hazardous, but it put American aircraft within striking distance of Vietnam.

The initial bombing runs sought to destroy supply lines on the Ho Chi Minh Trail in Laos. Then, in February 1965, after an attack on an American compound and helicopter base in South Vietnam, the targets moved to North Vietnam. The next month, the

15

Johnson administration launched the bombing campaign Operation Rolling Thunder, its name derived from a hymn; it would continue, with incremental expansions and occasional pauses, for three years. By April, the Air Force and Navy were flying 1,500 sorties a month against the North, which increased to 4,000 in September. In June, the United States had 75,000 combat troops in South Vietnam. By the end of the next month, that number had increased to 125,000.

Cherry knew little about the conflict except that he was fighting the Communists, that the South Vietnamese had a right to choose their own form of government, and that initially everything was very secret. The bombings were categorized as "classified missions," so classified that he didn't tell his wife, Shirley, who assumed he was still sitting on alert in Korea. (He eventually told her the truth.)

At the outset, using intelligence information, the squadron leaders identified a slew of targets, such as ammunition dumps, radar sites, airfields, bridges, industrial centers, power plants, and the flood control system of the Red River Delta. They mapped their routes and determined what bombs to use, but the plans were never implemented. Instead, they were ordered to blow up roads or mountainsides, sometimes to start a rockslide that would bury a passage. They were allowed to hit early detection radars, but those sites were soon removed from the target list. Cherry assumed the country's civilian leaders were choosing low-impact targets to avoid unnecessary destruction, but he also knew that this was no way to fight a war. During the Korean War, he bombed the Toksan Dam, flooding a valley to destroy bridges, highways, railroads, shelters, and an airfield. Civilians as well as combatants were drowned. It was horrible, but such attacks helped bring an end to the fighting.

In the early stages of the Vietnam War, Cherry knew the attacks were not crippling the enemy, and some of the assignments had a bizarre bureaucratic quality. In one case, the airmen were ordered to destroy a military complex near the city of Vinh, which it did by flying sixteen sorties (one sortie is one plane attacking one site). But Washington demanded that eighty sorties be flown, so the pilots had to fly another sixty-four — with nothing to bomb except rubble — to satisfy the order.

If the bombings seemed to produce meager results, they were by no means without risk. The enemy, Cherry discovered, had developed a sophisticated air defense system, which was shooting down American aircraft at alarming rates. The system itself featured surface-to-air missiles, antiaircraft cannons, a complex radar system, and computerized control centers, all provided by the Soviet Union. The city of Hanoi had the most formidable air defense in the history of warfare, though no place in the North was truly safe. Automatic and semiautomatic weapons were passed out to anyone in the countryside who could shoot at a plane. After Rolling Thunder's first six months, more than thirty airmen had been killed or were presumed missing; a dozen had been captured. Fred Cherry was almost among them. On one treetop attack, he encountered a barrage of small-arms fire. With his eyes closed and sweating through his flight suit, he somehow pulled up safely. He knew he was lucky to have survived, but for now he retained that special feeling, that mojo, that emboldens every fighter pilot. He was still invincible.

Using air power in driblets was part of America's strategy of fighting a "limited war." The attacks were meant to prod the North to the negotiating table, where U.S. interests — the preservation of South Vietnam's anti-Communist government — would be ensured. They were also meant to weaken the morale of the North's leaders, who might then call off the Southern insurgents. Fearing a broader conflict against China, U.S. officials believed they could win a "limited war" through a "graduated response" of military force.

But they underestimated the resilience of the North Vietnamese, who for two thousand years had been fighting foreign invaders — the Chinese, the French, the Japanese — as well as North Vietnam's willingness to endure devastating losses (three million people were killed in the war). To the Communists, the battle against the United States was a continuation of their battle against French imperialism — both were wars of attrition that the outsiders could not win. As Ho Chi Minh, who led the crusade, told the French on the eve of their colonial war in 1946: "You can kill ten of my men for every one I kill of yours, but even at those odds, I will win and you will lose." Whatever the outcome of the American war, U.S. casualties would be significant, and pilots were among the most vulnerable.

Cherry completed forty-six combat missions after three tours in Thailand. He was then told that his time in Asia was over as part of the normal rotation of pilots. In October 1965 he was given two weeks to return to the United States, where he would instruct pilots on the F-105 at McConnell Air Force Base in Wichita, Kansas. Cherry enjoyed teaching — it maximized his time in the air — but did not want to leave Asia. He was still optimistic that the war would be won and that the bombing missions, however hamstrung, were damaging the enemy. He also assumed that, at the age of thirty-seven, if he returned to the United States, he would never see combat again.

Cherry wanted to accompany his squadron on its next tour to Thailand, which was to leave on October 18. The night before its departure, he drove across the Yokota Air Base, near Tokyo, and stopped at the Yokota Officers' Club. He had one night to convince his commander that he should stay.

The club itself, a one-story building, was once an American serviceman's paradise. In the 1950s, a lighted porte-cochere welcomed visitors, banquettes were covered in silk, and rattan furniture filled the Samurai Ballroom, scented with cigar smoke, perfume, filet mignons, and spiced red apple rings. By the middle 1960s, the club had lost some of its glamour. Gone were the exotic appointments, replaced by molded plastic furniture and Formica tables. The rugs were worn, the paint faded, the jukebox old. But the club was still the social hub for the entire complex of American operations officers, "ground pounders," paper pushers, and desk jockeys, while slot machines, live music, and rivers of Scotch provided the entertainment.

Cherry knew that his commander, Lieutenant Colonel William Peters, would be there. A ruddy-faced officer nicknamed Napoleon, Peters thought highly of Cherry, having recently recommended him for an air medal for his Vietnam missions. He also endorsed Cherry's promotion from captain to major. But when Cherry found him at the bar and asked to prolong his stay in Japan, Peters waved him off.

"Fred, you've done your job in Southeast Asia," he said, fingers wrapped around his glass. "Get your ass back home. You've got a job to do there."

Cherry pressed his case, arguing that he might never fly in combat again. "This is my last chance," he said.

They argued for several minutes until Peters finally relented. "Oh, goddamn it, Fred, go ahead," he said. "But get your damn ass back here in two weeks."

Cherry celebrated the news by meeting up with a close friend, Marvin Walls, a captain for the Reconnaissance Technical Squadron, which identified targets in photographs. Sitting at the L-shaped counter in the main bar, a mirrored wall lined with beer bottles, Cherry and Walls drank Scotch and toasted Fred's good fortune until the bar closed at 2 A.M. Then they headed outside, and when Walls began walking down the steps, Cherry reached out and tapped his shoulder. Walls, five inches taller, stood several steps down from him, looked into his eyes, and saw something odd — not fear, but resignation. "I don't know," Cherry said. "I have a funny feeling about this. I don't know if I'm going to make it."

Walls had always known Cherry as cocksure; now, speechless, he embraced his friend.

Cherry didn't know what caused his sudden reversal of confidence or why he would even disclose his premonition. The moment would haunt him for years to come, but he had no time for fears. His wish for one more round of combat had been granted.

3

On Target

On the morning of October 22, Fred Cherry, his flight suit unzipped, attended a predawn briefing at the Tactical Operation Center of the Takhli air base. It was overcast and warm, and a fan hummed quietly. Cherry sat with four or five other airmen in what would be a highly unusual briefing. On a typical flight, Cherry would be assigned a target and be given at least half a day for preparation, learning the correct compass headings, air speeds, altitudes, the call signs of other planes, and a dozen other details. He would be responsible for drawing his own map, sketching in rivers, mountain ranges, railroads, and other navigation markers and creating the actual route. He would also have multiple photographs of the target, each from a different angle and distance, which would help him chart his route. Finally, he would typically fly in good weather.

None of those conditions existed that morning. Intelligence officers handed him his map and his route and said he'd be leaving in several hours. He was given only one photograph of the target, taken immediately above the site. The expected rain had canceled all other flights. The hastily conceived attack was formed after the discovery of a surface-to-air missile installation fifteen miles northeast of Hanoi.

A senior officer identified the installation on a map plastered on the wall, saying, "We have to knock it out as soon as we can."

A surface-to-air missile, or SAM, was a large munition shaped like a telephone pole and launched like a Roman candle, with wings attached to the side, fire blowing out the back, explosives packed inside, and radar guiding its trajectory. It was a devastating counterpunch to America's air attack, able to knock a plane down at 50,000 feet.

The first victim of a SAM was Air Force Captain Richard Keirn, who was shot down on July 24, 1965, and taken into captivity.* Thereafter, when a SAM site was discovered, the United States immediately ordered a retaliatory bombing raid, known as an Iron Hand. Ironically, these missions to eliminate a dangerous weapon were perhaps even more dangerous than the weapon itself. Because a SAM's radar had difficulty tracking aircraft below three thousand feet, U.S. fighters flew low. In response, the North deployed gunmen with antiaircraft weapons — cannons — to protect the sites. It also created dummy (fake) installations, luring pilots into an ambush. On a single day in July, antiaircraft fire shot down six planes on Iron Hand missions.

Cherry knew the risks of that day's mission. The low cloud cover gave him a two-hundred-foot ceiling, meaning he could avoid small-arms fire only by flying above the clouds. But that was impossible, as the clouds would have shielded him from his navigation markers and his target. The bad weather effectively left him more exposed. His route was also a concern. It forced him to follow the Northeast Railroad, the main supply line from China to Hanoi. While it may have been the most direct route, Cherry

*Known as "Pop" Keirn, the forty-one-year-old was captured near the end of World War II as well and spent seven months as a Nazi prisoner.

22

knew that it was heavily defended by antiaircraft weapons and that he was likely to draw fire.

The briefing itself lasted less than ten minutes, and Cherry then stayed in the command center to study some maps. A senior officer with eagles on his shirt collar approached him — Colonel Shook. Several days earlier he had been introduced to Cherry on the airman's arrival in Takhli. Though Cherry was wearing his flight suit, the colonel refused to accept his status, saying, "Well, duty officers don't fly."

Accompanying Cherry was a lieutenant colonel, who immediately corrected Shook, but Cherry believed the snub was racist. His flight suit made his status clear, but as the only black combat pilot in Takhli, he was not easily accepted by everyone.

That day, as Cherry pored over the maps, Shook approached him. "What's the matter, boy?" he asked. "You're not up to it?"

Cherry was incensed but helpless. "Don't worry," he told the colonel. "I'll do my damn job." He stood up, grabbed the maps, and left the room.

Watching the incident was First Lieutenant Bruce Rankin, a pilot who would later wonder how the colonel's actions may have influenced the day's events. "The colonel may have denied Fred the opportunity to make a better flight plan," he later said. "It wasn't fair to Fred." Cherry himself would always regret that he didn't have a few more minutes to study his maps.

Cherry was supposed to take the afternoon mission; another pilot, Captain Michael Cooper, was to be a flight leader for the morning assignment. After the briefing, Cooper returned to his hooch. "If they launch me, give me a call," he told Cherry. "I'm going to take a nap."

Cherry was afraid the afternoon flight would be canceled and he would be denied a day of flying, so he told Cooper not to worry. "I'll take your line, and you cover me in the afternoon," he said. Despite his misgivings about the mission, he still felt invulnerable. Cooper, who believed the mission was reckless, didn't object.

Cherry was in his hooch when his call came. As he opened the door to leave, he stopped, returned to his desk, and removed his wallet, a loose credit card, and a pen with a U.S. government insignia. According to Air Force rules, a pilot was to carry only a military identification card, a dog tag, and a Geneva Convention card. If he was captured, the enemy could use personal information against him. Cherry had always ignored these requirements, but that day he felt different. He left his things behind.

At the flight line, he put on his antigravity suit. Resembling a pair of zippered chaps, it inflates during tight turns to prevent a sudden blood rush that could cause a loss of consciousness. He also wore his combat vest with a radio and battery, a .38 revolver, a hunting knife, flares, iodine pills to purify water, and diarrhea pills. His blue and white helmet and shaded visor came next. He secured his parachute when he was in the cockpit.

On the flight line, the pilots spun their engines in one plane after the next, filling the air with black smoke and creating a roar that was literally deafening. (Many pilots suffered some degree of hearing loss.) The F-105s had already been checked by the flight crew, but Cherry, as required, walked around the silver plane to inspect for leaks, foreign matter in the engine, or any other problem. The destructive power of a jet fighter was familiar and comforting: more than a hundred cluster bombs with pineapple wings were nestled in gray canisters beneath the wings; a 20-millimeter cannon, which fired six thousand rounds a minute, was perched on the plane's nose.

A young corporal checked the jet's weapons and gave Cherry a snappy salute. "You're loaded to the teeth, sir." Cherry saluted back.

He usually flew the same plane, and he worked with the technicians to keep his radar "perked," or clear, and to ensure that his computerized weapon system was highly tuned. While many pilots ignored the automated bombing system as too complicated, Cherry didn't. He believed it narrowed his margin for error, for it calculated the distance to the target and the optimum angle at which to release the weapons.

Inside his cockpit, he flipped on the starting switch and gave his wingman and the two other fliers the signal, rotating his forefinger in circles, to ignite their engines as well. He used hand signals to communicate to the ground crew but also had a microphone in his oxygen mask.

"Hot and muggy," the air traffic control officer told him.

Five minutes passed before the plane moved. When the control officer said, "Clear to taxi," Cherry checked with all the flight members and gave them a thumbs-up. The crew chief pulled the chocks from the wheels of all four planes, which began to taxi toward the runway. Cherry pulled down and locked his plexiglass canopy. The crew chief saluted each pilot, who responded in kind. Then, right before takeoff, a problem arose. A wingman's plane, the jet that would fly in close tandem with Cherry's, experienced a sudden loss of oil pressure and had to abort. A new aircraft would be used, but it didn't have the same weapons as Cherry's. To coordinate the attack — to ensure that the weapons fell on the target simultaneously — the wingman would have to fly ahead of instead of adjacent to Cherry. It made no practical difference for the flight, but any deviation from routine was disconcerting.

Finally, they were ready for takeoff. Cherry's jet shook slightly as it powered up, then sped down the runway and lifted quickly into the gray sky, the force pushing him hard against the back of his seat. On some takeoffs, the pilot was blinded by the sun caroming off flooded rice paddies. But today Cherry saw only verdant jungles rush by below. He broke through the clouds and saw the sun. He checked his course settings and, cruising at 680 mph, relaxed. It was ten-thirty A.M.

The radio crackled in Cherry's ear, and bits of conversation from distant ships drifted in and faded quickly. They had left Thailand's airspace and were now over Laos at the 19th parallel, where they would refuel. Flying above the clouds, everything seemed smooth. Cherry heard his code name on the radio. The KC-135 tankers, which had been orbiting, were now approaching. He scanned the sky briefly and found them. Matching speeds, he flew behind one tanker, whose boom, or long tube, nudged its way into the Thunderchief's external gas tank, and in three minutes the tanker furiously pumped in one thousand gallons of jet fuel. It was a routine job, but when Cherry was hooked to the fueler twenty-five thousand feet above the ground, he always felt more like a target. When the task was finished, the boom retracted; Cherry rolled away, and his wingman followed in line. In less than fifteen minutes, all four aircraft had been refueled.

He flew above the clouds until they reached the mountains near Dien Bien Phu; heading east, he rocked his wings to signal his wingman, then descended sharply, slicing through the clouds and leveling off above the trees at a hundred feet. He was now "on deck," flying at 575 mph, low enough to avoid enemy radars and to feel the full rush of flying. At high altitudes, a pilot barely feels as though he's moving; at treetop level, he has a complete "awareness of speed," as Cherry called it, which conferred an ever greater feeling of power, even omnipotence. He was thirty-four minutes away from the site.

He raced ahead until he reached Kep airfield, then turned southwest, which would put him parallel to the Northeast Railroad. The airfield, however, was fortified with antiaircraft weapons, and as Cherry passed by, gunmen fired tracers that looked like flaming orange tennis balls. They narrowly missed him, and Cherry accelerated to 700 mph, still flying straight and low.

The dangers had just begun. Three minutes from his target, Cherry saw hundreds of rifles pointed at him, their muzzles flashing. Combat pilots liked to "fly and fight," vacating battle sites quickly and avoiding protracted gunfire, but now Cherry faced just that. As he later discovered, armed peasants had been rebuilding a road that had been destroyed by a U.S. bomber, and he assumed they had been alerted to his raid by the gunmen at the airfield.

There was no point in deviating from course. He was right on top of the riflemen, and to pull away would give them an easier shot, so he gripped the control stick and searched for the target.

Thump! Metal slammed against metal. His plane shook and swerved. "I've been hit," he yelled into his microphone.

He didn't know where the damage was, but the computer system on his flight control panel shut down, destabilizing the aircraft and leaving him in sporadic radio contact with his wingman. He locked the control stick between his legs and used both hands to steady the jet, but it jerked and yawed. Still, he maintained altitude and control. His mouth and throat were dry from the near-pure oxygen he had been breathing through his mask, and perspiration soaked his combat suit. He turned off the electrical and hydraulic switches to minimize the chance of fire. Cherry saw a lake at a bend in the railroad, which was supposed to be near the missiles. He finally saw the installation — several battery launchers that formed a circular pattern. His pellet-spraying cluster bombs were "antipersonnel weapons," designed to remove any militiamen so that his wingman and the two other fliers could bomb the site without fear of counterattack. When he reached the target, he held the red button on the control stick for three seconds, which emptied the bombs from their canister. Through his rearview mirror, Cherry watched them hit the installation, exploding in little balls of fire.

He then flew away from the site and felt the plane straining skyward. His plan was to fly about forty miles to the Gulf of Tonkin, where he would ditch the Thunderchief and be rescued by a Navy carrier. "Let's get the fuck out of here!" he yelled to his wingman.

But then smoke began to pour from his instrument panel and warning lights dotted the cockpit like a Christmas tree. The plane shook violently, and Cherry, losing control, gripped the stick tightly and tried to straighten it out. The clock read 11:44 A.M. The cockpit continued to fill with thick smoke. He had one last chance to save the aircraft. He leaned over to turn off the switches for the battery, generator, and alternator, believing he might minimize the fire while keeping the engine alive. But Cherry's hand never reached the panel. A loud explosion sent the Thunderchief spinning through the air. The blast likely came from the 20-millimeter ammunition in the aircraft's nose — allowing Cherry to remark later that the North Vietnamese didn't shoot him out of the sky. He shot himself.

The plane was going to either explode completely or crash in seconds. Cherry pulled on the control stick, tilting the jet's nose for ejection. At about 400 feet — high enough to bail out but low enough to minimize the exposure to gunfire — Cherry pulled the ejection handle with his left hand. The instrument panel shattered as the canopy flew off, creating a wind tunnel effect. Cherry's left arm, unsecured, was sucked straight up and wrenched from its socket. Then he was literally shot out of the aircraft by a 37-millimeter cannon shell from beneath the seat. It is supposed to create a "smooth ballistic trajectory," in which the lap belt disengages, the parachute automatically opens, and the seat falls away. But none of that happened. His lap belt didn't unlock and his chute didn't budge. He remained strapped to his seat, sailing harmlessly over the Vietnamese brush. The malfunction probably saved his life. The maximum speed at which one can release a parachute safely is about 575 mph, and Cherry was flying at close to 700 mph; the chute would have ripped apart.

When Cherry's eyes were able to focus, he saw the sky and realized he was still in his seat. Tilting back, he pulled the ripcord. The white canopy blossomed above, and he finally disengaged the seat. As the ground rushed toward him, he looked at the chute's fluffy panels, saw the dark form of his wingman fly by, and heard bullets whiz past his ear. Just before impact, he saw black and gray clouds drift over the ridge where his plane had crashed. He slammed hard against the high grass on a small hill. He was two minutes from the coast.

"Our lead got out," his wingman radioed to the base in Thailand. "We saw him hit the ground, but I don't think he was conscious."

He was conscious but badly injured. In addition to ripping out his shoulder, the fall had broken his left wrist and left ankle. Even healthy, he wouldn't have escaped: he was immediately surrounded by a dozen armed militia, as well as a bunch of kids with hoes and pitchforks.

"Damn," Cherry said under his breath, "I'll be here a long time." He assumed it would be one or two months before the United States won the war. He was the forty-third American captured in North Vietnam, and the first black.

4

Hanoi's Welcome

A young pilot in Cherry's squadron used to say that he would rather die in combat than be captured: death was immediate; captivity was long, excruciating, and sometimes fatal. Cherry told the pilot that his choice was foolish. "You can always die," he said. "But you at least have to go in and test the waters."

Cherry was now living out his own advice, yet there was something incomprehensible about his position: moments earlier, he had been piloting the largest single-seat, single-engine fighter ever built, a twentieth-century warrior catered to by enlisted men, support personnel, and military gofers. Now he was being mocked by giggling children with farm tools. Cherry wasn't afraid; he was just dumbfounded.

A militiaman with a semiautomatic weapon stepped forward and showed two fingers, indicating he wanted Cherry to raise both hands. The American wasn't in pain yet, but he couldn't move his left arm. His .38 pistol was in its harness, the butt showing. Cherry feared that if a militiaman saw it, he would be shot. He used his good hand to motion to his gun, and two soldiers finally moved in slowly and took it, as well as his hunting knife. They also took his parachute and antigravity suit; they wanted to remove his flight suit as well but were baffled by the zippers. A Vietnamese civilian took Cherry's knife, walked over to him, and plunged it toward his groin. The American scooted backward, and the knife sank into the ground inches from his genitals. Cherry calmly demonstrated the zipper. Amused, the man began toying with it — up, down, up, down. He got the flight suit, but when he indicated that he wanted to cut off Cherry's boots, the American resisted by kicking. He didn't want to walk barefoot, and there was something special, even sacred, about a serviceman's shoes. After a brief scuffle, whoever was in charge told the man with the knife to let the pilot keep his boots.

A militiaman motioned for Cherry to stand, so he rocked forward and picked himself up, dusty, hot, and sweating, his face nicked and bleeding from the instrument panel's shattered glass. His left ankle hurt, but he didn't know it was broken. Someone pushed him with a stick and, escorted by his captors, he limped to a path and headed for a nearby village. He had heard the Vietnamese language before and even understood a few words, but he didn't understand his captors as they talked and laughed among themselves.

It was a two-mile walk, and he soon grew tired, shambling along the path as best he could. As they approached the village, a gong sounded, the loud rumble beckoning farmers from the rice paddies. Children ran his way and touched him. He tried to smile. They laughed. A young man in uniform seemed to be in charge, which heartened Cherry. He assumed that a soldier, even a Communist, was more likely to respect a prisoner of war. According to the Geneva Conventions of 1949 — which North Vietnam had signed — POWs were to be treated humanely.

Cherry, at this point, was not in danger. Ho Chi Minh had already announced that any village capturing an American would be rewarded if it returned the prisoner alive to the authorities. To

31

his captors, therefore, Cherry was more a prize than he was the enemy. The militiamen took him into a hut, where a medic put ointment on his face to soothe the cuts while passersby looked in. When it was time to move again, precautions were taken. The soldiers put a black cloth over Cherry's white shirt so he'd be more difficult to spot from the air. His ankle was also swelling, and he finally lost his composure when they were walking out of the village and a civilian's bike rode over his feet. The American grabbed the handlebars and shoved the bike over a hedgerow into a rice paddy, the rider in tow. The man, furious, charged the pilot, but the soldiers blocked his path and turned him away. Cherry's elbows were soon tied behind his back, stretching his dislocated shoulder.

The American realized that he was valuable to his captors, but moments later he thought his life was about to end. Two U.S. jets appeared from behind a mountain and flew low, searching for the downed pilot by homing in on the electronic signal in his parachute. The soldiers threw Cherry face down into a dry rice paddy, and one man straddled his back, pointing his automatic weapon flush against the bone behind his ear. Cherry, his nose in the dirt, his shoulder aching, assumed that if a jet pilot saw him and took action, he'd be dead. But the planes passed by without finding him. The gunman relaxed his weapon, picked up his captive, and pushed him forward.

As Cherry limped to a Jeep, a man with a camera wanted to photograph him. Such pictures, he rightly assumed, would be published or displayed to boost the pride of the Communists, so he turned his body, fell to one knee, frowned, and refused to give the cameraman a good look. The photo session soon ended without a single picture; Cherry counted it as one small victory. They piled into the vehicle and took off.

The ride relieved his ankle, but with his elbows tied, he couldn't lean back. Angry, he nudged a guard, saying, "Too tight, too tight!

Hurt, hurt, hurt!" He asked him to loosen the nylon cord, but the guard refused. Cherry, confident the men wouldn't shoot him, knocked the guard with his shoulder and almost pushed him out. The soldiers in front laughed, discussed what to do, and finally allowed the guard to loosen the rope.

They drove to yet another village and stopped at a brick building surrounded by a courtyard jammed with teenagers. It appeared they were now at a school, and when Cherry entered the structure, he feared his journey had taken an ominous turn. He sat down at a table covered by a blue cloth, an interrogator across from him, a guard with an automatic weapon behind him. The questions — Who are you? What aircraft were you flying? What were your targets? — were in English.

Each response was the same.

"Fred Cherry. Major," he said. He gave his serial number and date of birth, and he took out his Geneva Convention Identification Card, which outlined his rights as a prisoner.

"Forget about it," an interrogator said, ripping it up. "You're a criminal."

Cherry could hear hundreds of teenage students chanting in Vietnamese. The interrogator said, "They're yelling, 'Kill the Yankee!'"

"So kill me," Cherry said. "You're going to do that sooner or later."

"You are a criminal," the interrogator said.

"I'm Fred Cherry. Major."

After a few more fruitless exchanges, a guard bound Cherry's elbows again and they headed outside. They passed men and women in peasant clothes and conical hats, some carrying hoes, shovels, and other hand tools. Cherry's presence was a diversion from harvesting. Suddenly, a young farmer ran toward him and rubbed his hand. The militiamen, fearing an attack on the American, pushed their rifle bolts forward. But there was no attack; the peasant just wanted to see if Cherry's color rubbed off. He quickly disappeared back into the crowd.

Cherry was returned to the Jeep, and they drove all afternoon and into the evening. Almost all vehicles were heavily camouflaged. Headlights were seldom used, but when they were, they showed only through the bottom third of each lens. (The rest was painted over.) Cherry had not had food or water since the morning, but his exhaustion was greater than his hunger or thirst. He just wanted to lie down.

By ten P.M. they reached Hanoi, one of Asia's oldest cities, where pastel villas and spacious verandas recalled the French colonial rule. The French left something else as well: the Hanoi Hilton. It was Fred Cherry's next stop.

There is never a good time to become a POW, but Fred Cherry's timing could not have been worse. His arrival coincided with the North's first crackdown on American prisoners.

Navy Lieutenant (j.g.) Everett Alvarez, Jr., was the first American POW in the North, captured in August of 1964 after alleged attacks on U.S. destroyers in the Gulf of Tonkin. The Vietnamese had not developed confinement procedures — the Hoa Lo Prison still held civilian convicts — and Alvarez's treatment, initially, was bearable. The Johnson administration, gearing up for a presidential election, stopped the air raids, contributing to this relatively benign environment. While the food was bad, Alvarez was able to write and receive letters, to leave his cell for exercise, and even to read. Loneliness was his biggest enemy. Using a nail, he marked the passing days by scratching messages in one corner of the prison: "Happy Labor Day." "Have a good Thanksgiving." "Merry Christmas." He suffered daily interrogations, but he found his examiners more amateurish than cruel. They lectured him about

Vietnam's history of wars against various oppressors and exhorted him to write a letter to Ho Chi Minh expressing his appreciation for his favorable treatment. He refused.

As the number of POWs increased, they were routinely subjected to questioning, bullying, and indoctrination. Political officers hectored them about the evils of capitalism; demanded that the Americans denounce their government and acknowledge their own criminality; occasionally threatened execution; and inflicted some physical abuse. If a prisoner was caught violating any rule, such as communicating to another inmate, he was locked in isolation. Life was hard, but not horrific.

The situation changed in the fall of 1965, which happened to be one of Hanoi's coldest autumns in years, forcing America's three dozen prisoners to shiver in the concrete chill. But the weather was the least of their problems. The Vietnamese officials abruptly escalated their physical abuse, using torture as almost a rite of passage for virtually every POW in the North.

The change came about for several reasons. On October 24, the guards raided the prison cell of Air Force Lieutenant Colonel Robinson Risner, an American hero even before he was shot down. He had been heralded in a *Time* cover story as an example of the dedicated American fighting man in Vietnam. Unfortunately, the enemy saw the story — it received copies of America's leading magazines — and made Risner's life miserable. When guards ransacked his cell, they discovered a list of directives that the Americans had been passing around, explaining how they were to create a chain of command, communicate among themselves, and frustrate their captors. The guards also found an iron bar that was used to drill holes in the wall. To the Vietnamese, these were the first pieces of evidence that the Americans were organizing to resist, in essence to continue the war behind bars. Reprisal was swift. Rules were enforced with torture, resisters were

35

punished ruthlessly, and privileges were denied — Red Cross packages, for example, were seized. Even the food was downgraded. Until then, the POWs had gotten one and sometimes two bananas a day. After the crackdown, months passed before some prisoners saw fruit again.

James Stockdale, a Navy commander who was captured a month before the crackdown, later wrote: "By carrying out a new policy action, North Vietnam had crossed a boundary. Henceforth, Americans were to be allowed to stay within the bounds of name, rank, serial number, and date of birth only at North Vietnam's sufferance." The prisoners later dubbed this campaign, lasting from late October 1965 to the fall of 1969, "the Extortion Era."

Even without discovering contraband, the Vietnamese may still have tightened the screws. America's frequent bombing raids, while not crippling the country, were damaging. The North realized that hostilities could persist indefinitely — as could the prisoners' captivity — so imposing strict discipline on them would diminish their resistance. More important, the North could exploit the Americans to advance its own goals. In past wars, combatants used their POWs to negotiate more favorable peace terms, but the Vietnamese believed they could use their prisoners to elicit — through whatever means necessary — statements condemning the U.S. government. Borrowing interrogation and propaganda techniques from the Chinese and Korean Communists, they believed such pronouncements, written, taped, or filmed, would boost the morale of their own people and stir up antiwar sentiment in America. As a Vietnamese official told Stockdale, "Our country has no capability to defeat you on the battlefield. But war is not decided by weapons so much as national will. Once the American people understand this war, they will have no interest in pursuing it . . . We will win this war on the streets of New York."

The Vietnamese recognized the importance of public opinion

in America but were remarkably naïve in believing that a statement from a POW — or anyone in captivity — would carry any weight. Nevertheless, they were never stymied by their lack of sophistication. Torture would be the tool to extract POWs' statements, to break their will. Many prisoners believed that torture was an extension of Communist doctrine, giving rise to the definition "A Communist is a person who will torture you to write a statement that you are not being tortured."

But the roots of such abuse lay deeper. The Hoa Lo Prison had been built by the French for their Vietnamese captives. The leg irons, manacles, and handcuffs that fit snugly on the small frames of the Asian prisoners cut deeply into the flesh of the Americans. Ho Chi Minh, in *French Colonialism on Trial* (1926), condemned torture as a means of oppression, but after he assumed power, he used it to crush opponents. Ho's chief military strategist, General Giap, said in 1956: "We . . . executed too many honest people . . . and seeing enemies everywhere, resorted to terror, which became far too widespread . . . Worse still, torture came to be regarded as a normal practice."

It was normal practice for the pro-American government in South Vietnam as well. Its regular police and security agencies, trained by the French, tortured suspected Communists to pry out the names of other cadres, then either shot them or sentenced them to a concentration camp. In the infamous "tiger cages," prisoners were beaten with a bludgeon or an electric whip. Women arrested were usually raped as well as tortured, because, as the American journalist Neil Sheehan wrote, "The torturers considered rape a perquisite of their job."

Vietnam's history of torture made its use against the American POWs all but inevitable; it was simply bad luck that the prisoners captured at the end of October 1965 arrived with its most ruthless application.

*   *   *

The Jeep carrying Fred Cherry drove through Hoa Lo's big front gate. The vehicle cut through an outer stone wall, rumbled across a cobblestone alley, turned into a tunnel, passed through another set of gates, and then stopped in a courtyard. The distance between the street and the courtyard was only about seventy feet, but the clanging of the metal gates and the eerie blackness of the tunnel signaled a passage into a forbidding world.

Built at the turn of the twentieth century with a capacity of two thousand offenders, Hoa Lo was North Vietnam's main penitentiary and the headquarters of the country's entire prison system. Occupying an entire block, it was surrounded by a concrete wall about sixteen feet high and six feet thick. Embedded in the top of the wall were shards of iridescent greenish-blue glass, said to be the remnants of French champagne bottles. Beyond the glass were three strands of barbed wire, one of which was electrified. Guard towers stood at the prison's four corners.

The courtyard itself had a veneer of order and serenity. A bit larger than a basketball court, it was lined with faded, two-story white stucco buildings with red tile, which served as administrative offices. Along the cobblestone driveway stood four well-tended flower beds for the pleasure of prison officials or visitors — but not the inmates, who were delivered to a nearby cellblock.

Cherry barely noticed his surroundings. Fighting through sleep, he worried whether his wife and children knew he was alive and hoped that they could get back to the United States and find someplace to live. He was taken to an area that the Americans had dubbed the New Guy Village, where the Vietnamese inflicted their worse torture. Room 18 had soundproofed walls and an array of menacing contraptions, including a giant hook suspended from the ceiling. Ropes were tied around the inmates' arms and strung up on the hook, the cord sometimes soaked with gasoline to intensify the pain. Catty-corner to that was a second chamber, the "knobby" room, whose pale green walls were covered with rough knobs of acoustic tile that muffled screams, the tile cracked from the impact of many bodies.

Cherry walked into that room and recognized a familiar arrangement: the rickety wooden stool, the desk covered by a blue cloth, the interrogator, the guard. A conical shade on an overhead light bulb could be used to direct light into the eyes of the prisoner. A tape recorder hidden beneath the table could capture and later broadcast damning statements.

His arms still tied behind his back, Cherry sat so close to the desk that he could barely move his knees. By now he was so tired that he struggled to keep his head up and his eyes open. The pain in his left shoulder was increasing, but, fearing the injury would be exploited, he did not acknowledge it to his captors. To stay awake, he needed a diversion, and he noticed a plant in a large urn against a wall. The plant drifted in and out of focus, but Cherry kept his mind on it.

His interrogator, the man called Rabbit, sat before him while a guard stood behind him.

"You're a criminal," Rabbit said. "You committed crimes against the Vietnamese people. Are you going to admit your crimes?"

Cherry shook his head.

The guard whacked him across the head with his palm, then kicked his chair out from under him, the pain shooting through his body as he fell to the floor. He was lifted up and placed back on the chair, and he looked at Rabbit again.

"What was your mission? Who was in your squadron?"

Cherry stated his name, rank, serial number, and date of birth. He tried to brace his body before the next assault, but to little effect. The guard grabbed his head and crashed it against the table, then kicked the chair out.

In seconds he was back on the chair, and the questions continued.

"You killed thirty people," Rabbit said, his voice rising. "Do you feel good about that?"

This time Cherry responded. "I didn't kill anyone," he said. "I just gather information."

"What kind of plane did you fly?"

"An RF-105 reconnaissance plane," Cherry said. The Air Force didn't have such a plane.

The officer wrote the information down, then looked up and said, "You have no RF-105s."

"We sure do," Cherry said. "I had one."

"How does it work?"

"I have no idea. The pilot just pushes the button, leaves it on, then turns it off. That's all the pilot knows. Okay?"

Months later, Cherry saw in a Vietnamese magazine a photograph of the tail of an F-105, and it was called an RF-105. He figured it was his plane because he doubted anyone else used that lie.

A hand grabbed the back of Cherry's head and again slammed it to the table.

"What was your destination?" Rabbit spread maps out on the table, and Cherry saw they were his, retrieved from his downed Thunderchief.

Rabbit continued asking questions, but Cherry was too groggy to hear much. Throughout the night, he was knocked to the floor, was punched across the ears and neck, and had his head slammed against the desk. Blood trickled out of his nose, welts formed over his eyes, and his ears rang. At one point he murmured, "Under the Geneva Conventions . . . you can't treat a prisoner this way."

"You are not a prisoner of war!" Rabbit stood up quickly. "You are a criminal!"

Cherry finally realized that his POW status meant nothing because no one, with the exception of the enemy, even knew he was a POW. The enemy could kill him, and his death would be attributed to the shootdown — killed in action.

Yet he was learning to weather the abuse. He relaxed his entire body and tried to think of something pleasant — flying combat missions. He envisioned maneuvering through the air, dropping bombs, eluding the enemy. His trancelike state gave way to actual sleep, and his head dropped, but a guard yanked it up and banged it on the table. As the night wore on, they continued to ask him about his plane, his missions, his targets, and what he knew of North Vietnam's defenses. He refused to answer their questions. At one point, his arms were twisted behind his back and pushed upward, further ripping the socket of his left shoulder. They tied his arms, milled about, chatted, drank tea. Had he yelled out, the knobby room would have muffled the sound, but he never screamed. Finally he blacked out, but he had survived his first day as a prisoner of war.

Just before dawn, he awoke and was taken to room 24, a cell with a concrete floor and nothing else, save the gray rats with webbed feet, scorpions, and ants. The doors were large and seemed solid except for holes in the bottom. He slept on the floor and was allowed to go into a small yard to relieve himself and to wash. He continued to receive daily interrogations. By then the Vietnamese recognized that his swollen ankle and contorted shoulder were badly injured, and they used that against him.

"If you don't cooperate, you don't see doctor," an interrogator told him. Cherry refused, so was denied medical attention. He could not shower or bathe; he would receive food twice a day, salty fat pork, moldy bread, and a few greens, or "swamp weed," as the Americans called them. Insufficient water left him dehydrated. The prison gong, sounded by a metal pipe against a railroad iron, dictated the daily regimen: a gong to wake up at six A.M., a gong for food at ten A.M., a gong for a nap at eleven A.M., a gong for dinner at four P.M., a gong for bedtime at nine P.M.

Keeping time, for Cherry and all the prisoners, was an obsession. Some watched the movement of the sun through slats in the cell or listened for the distant chime of bells in Hanoi, while others tried to steal glances at a turnkey's watch. To create a calendar, Navy Lieutenant (j.g.) Ralph Gaither used string from his blanket and tied a knot for each day, leaving extra space to indicate a new month.

After a few days Cherry was given a mosquito net, which he considered a gift from God.

On November 1 he was taken to another part of the prison, a corridor with four small cells on either side. The Americans called it Heartbreak Hotel, and its conditions — decaying plaster walls, foul wastebuckets, odorous latrines — were as squalid as his previous cell. But at least he was given an olive prison uniform, a cotton blanket, underwear, a toothbrush, a water jug and cup, soap, three pieces of toilet paper (to last ten days), and a small waste bucket that doubled as a stool. He could also speak with other American captives and learn of their mistreatment. He noticed that someone had carved a matrix on the wall — five horizontal rows and five vertical rows — with a different letter in each square. He had no idea what it was but soon learned that it had been carved by the cell's previous occupant. His name was Porter Halyburton.

After several days in Heartbreak, he was inexplicably returned to the isolation of room 24. His shoulder, wrist, and ankle were becoming increasingly swollen and painful, but he received no treatment. At 135 pounds, he was already lean, but he was still losing weight. The days began to pass in a fog, and the interrogations began to subside as his deterioration continued.

On the evening of November 16, a guard entered his cell and

rotated his wrists — the signal to roll up your belongings and get dressed. Cherry was blindfolded and put in a Jeep. He prayed that wherever he was going would be better than Heartbreak.

He was taken a few miles southwest of Hanoi, near the village of Cu Loc, where a prison had opened two months earlier. It was the third prison used for the Americans, and as their numbers mounted, the North Vietnamese would ultimately incarcerate POWs in fifteen different camps, though some operated for less than a year.* The two main prisons, however, were Hoa Lo and Cu Loc, both of which held Americans until the end of the war.

On the surface, the two sites could not have been more different. At Hoa Lo, one prisoner later said, "You could hear the screams of fifty years." But Cu Loc, apparently a former French film studio that still had yellowing posters, damaged reels, and abandoned auditoriums, evoked an art colony. If Hoa Lo was an entrenched hub of steel and cement, Cu Loc was the quirky suburban upstart, with ducks, chickens, and other animals roaming the grounds.

But the interrogation, isolation, and oppression were the same. To transform Cu Loc into a prison, the Vietnamese erected brick walls in fourteen single-story buildings to create numerous cells. But the buildings were still in disrepair, their windows boarded up and their interiors filled with dirt, broken glass, insects, and rodents. Outside, separate toilet facilities were built. A wall was constructed around the perimeter of the camp, and sentry towers were installed.

The POWs initially called the compound Camp America, and with farm animals about, designated buildings as the Barn,

---

*American POWs in South Vietnam lived in jungle camps, where conditions were much worse.

Chicken Coop, Pigsty, and Stable. Many of the louvered French doors had holes that allowed the guards to peer inside, but sometimes the livestock meandered by and gazed in, which gave rise to the prison's permanent name: the Zoo. As one inmate said, "It's the first kind of place where the animals come and look at the people."

For two weeks Cherry lay alone in his cell, the only daylight or air filtered through cracks and gaps in the door and through a brick-sized air vent high on the wall. A single, naked, low-wattage light bulb hung from the ceiling and stayed on day and night.

A small blue box with a radio speaker piped in an endless stream of propaganda as the pain from his shoulder and wrist spread through his torso. He ate little and felt too weak to move. The premonition he'd had at the Yokota Officers' Club was playing itself out. His captors had given him no reason for hope. But he had faced adversity his entire life, and he wasn't giving up. He was confident he would survive. He just didn't know how.

5

The *Independence*

On the morning of May 10, 1965, the USS *Independence* sounded its long, bellowing horn and shoved off from the gray coastline of Norfolk, Virginia. Its mission was to steam across the Atlantic Ocean, around South Africa, and through the Indian Ocean to the South China Sea, where it would assume duties with the U.S. Seventh Fleet in the Pacific. The attack carrier, to be gone seven months, received a patriotic farewell. Children waved American flags while a band played "Auld Lang Syne." Dignitaries toured the ship, wives and girlfriends said tearful good-byes, and Miss Norfolk, in a sleeveless white dress, white gloves, and white floral headband, smiled for photographers. The Norfolk Chamber of Commerce gave the captain a silver Goodwill Cup; the port city, proud of its naval tradition, also gave the *Independence* another memento — a bomb inscribed with white paint: GREETINGS FROM THE PEOPLE OF NORFOLK TO THE VIET CONG.

The ship was eighty thousand tons of steel and metal, a gray, angle-decked war machine that hauled forty-five hundred men and eighty fighter jets. Such a vessel is known as "a city at sea," loaded not only with mechanics, engineers, and sailors to keep it running but also with doctors, dentists, postal clerks, printers, career counselors, legal assistants, and educators. The *Independence* even had a musical band composed of shipmates who had brought their instruments.

The civic metaphor was fitting, but it hardly captured the delirious energy, the unremitting clamor, the sheer life-and-death drama of the enterprise. The jet names — the Phantom, the Intruder, the Vigilante, and the Skyhawk — conveyed the threat they posed to a distant enemy; but the planes themselves, loaded with fuel, cluster bombs, heat-seeking missiles, and 20-millimeter ammunition, could imperil their American handlers as well. A single miscue, particularly on takeoff or landing, could saturate the flight deck in a cataract of metal and flame.

On takeoff, a jet taxis onto a catapult track as crewmen race about, signaling with their scarred hands, ducking under moving wings, and looking for cover. The fighter engine wails as a deck officer in a yellow shirt waves his right index finger over his head. The pilot salutes from the cockpit and the deck officer drops his hand. The aircraft screams down the catapult, red flames spewing from its afterburners and steam billowing from the track. It accelerates to more than 100 mph in 250 feet. Just as it reaches the edge of the ship, its nose tilts up, and the machine is flung toward the sky. A jet that fails to reach sufficient speed crashes into the ocean.

The blast from takeoff can knock crewmen to the ground; anyone not working on the plane tucks his hands under his armpits to protect them from the heat. As the last jet takes off — planes can launch, day or night, every thirty seconds from four different catapults — crewmen turn around and find the first aircraft of an incoming mission. In seconds, it hits the deck and accelerates, trying to hook one of four "arrest wires" stretched across the ground. (Accelerating, though seemingly counterintuitive, gives the plane speed to take off again if it misses the cables.) Once hooked, the wire pulls taut and stops the jet; if it breaks, it snaps across the deck and can cut through a crewman like a weed-eater. And if the plane fails to stop after it engages the wire, without enough speed to become airborne, it will slam into a barricade or tumble helplessly into the ocean.

The *Independence* suffered two violent accidents on its Pacific cruise. On July 20 a Vigilante jet, returning from a reconnaissance flight, broke its arrest wire, could not stop, and dribbled off the carrier ninety feet into the sea. Two aviators were killed. Later, a tank on a Phantom ruptured on takeoff, spraying the flight deck with four thousand pounds of fuel — which was then ignited by the plane's afterburner. Roaring flames devoured the next plane in line and spread into a compartment belowdeck. Sixteen men suffered burns or injuries; no one was killed. The Phantom flew safely to shore.

If the flight deck represented organized tumult, then frenetic clatter buffeted the rest of the ship. Helicopters whirled above while squealing elevators lifted jets from their hangar bays to the deck. Carts drove bombs and missiles through the ship. Rock music blared in the cafeteria and bunkrooms, where men slept in three-decker cots, the gray nozzle of an air conditioner humming from above. Doors clanked. Pipes groaned. Chains crashed. Twenty feet beneath the water line lay a metallic jungle of valves and gauges that jeered and squeaked. Then there was the steam — the hissing vapor that whipped the planes down the catapults, cleaned the clothes and dishes, and powered the engines at 30 knots across the sea. At night, taps was broadcast throughout the ship.

It was Navy Ensign Porter Alexander Halyburton's first cruise, and not one he had envisioned. He had thought his first trip might be to the Mediterranean, whose exotic ports — Naples, Barcelona, Beirut, Malta, Genoa — would have been romantic rendezvous for him and his wife, Marty (a nickname for Martha). But Haly-burton volunteered for the *Independence*. He had been a Navy officer for only fifteen months and decided he should do as he'd been trained: fly in the back seat of a fighter jet as an RIO, a radar intercept officer, responsible for navigating flights and identifying targets.

Halyburton was no warmonger. Known as Haly, he was interested in literature, poetry, and prayer as befit a gentleman warrior, giving him what one friend called "a rich inner life." Another friend thought it was easier to envision him covering a war as a journalist than fighting in one as an airman. In fact, Halyburton had no intention of making the Navy his career. Years earlier he had rejected a coveted appointment to the U.S. Naval Academy because he could not abide its rigid way of life. Instead, he attended Davidson College in Davidson, North Carolina, his hometown, and in 1963 graduated with a degree in English.

He considered becoming a journalist or, enjoying the camaraderie of academic life, perhaps working as a college fund-raiser. He thought he needed to go to graduate school, but uncertain about his career and pressed for money, he could not justify the cost of an advanced degree. The military was not so much an option as an inevitability. He assumed he would be drafted, for the United States, responding to Cold War tensions (the Berlin crisis, the Cuban missile crisis), foreign brushfires (Laos, Vietnam), and international commitments (NATO, South Korea, Japan), desperately needed conscripts. Halyburton preempted his draft notice and volunteered, hoping to receive a better assignment as an officer.

He had never thought about being an aviator until his senior year in college, when a former fraternity brother told him about his experience flying a McDonnell F-4 Phantom, an extremely fast (Mach 2), maneuverable aircraft with sophisticated electronics that enhanced its radar intercept and bombing capabilities. That sounded exciting to someone who grew up in a town that didn't even have an elevator. In the Navy, Halyburton could travel around the world on great ships, learn how to pilot high-tech aircraft, and avoid sleeping in a pup tent.

Halyburton was also a product of the South, where Confederate generals were revered, the martial spirit was celebrated, and young men were taught that serving their country in combat was noble. As a boy, he watched the Davidson College ROTC band, color guard, and honor drill march through town. His mother told him about a cousin, a World War II Navy corpsman, who was killed during the invasion of Okinawa while caring for a wounded Marine and was rewarded with a Congressional Medal of Honor. That impressed young Porter.

His naval flight training did not go exactly as planned. He wanted to be a pilot, but he failed an eye exam because his eyes were tired from his college finals. He passed the test on a second try, but by the time the paperwork cleared, he had begun training as a flight officer and saw no reason to retrain as a pilot. An RIO used geometric guidelines to map out where his pilot dropped bombs and fired missiles, a kind of mathematics puzzle that Halyburton enjoyed. He also assumed his military career would be short. In February 1964 he was commissioned as an officer, receiving his shoulder boards with one gold stripe. When he left Norfolk in May 1965, he was obligated to serve twenty-six more months — ample time to return and take a Mediterranean cruise. He knew that many seamen back home feared the Vietnam War would be over before they had an opportunity to fight. For now, he was grateful to get at least one chance at combat.

Halyburton was awestruck by the *Independence*, where he was part of the VF 84 Squadron (the *V* stood for "fixed wing"; the *F* for "fighter"). The ship hauled more people than the population of Davidson and seemed to offer more diversions, sponsoring boxing matches, basketball games, skeet shooting, and variety shows. While Davidson's movie theater had burned down in the 1950s, the *Independence* showed a different film almost every night. Halyburton, who as the designated "popcorn officer" ensured that munchies were available, wrote to Marty about *The Sound of Music* and *West Side Story*, neglecting to mention the more popular pornographic flicks.

At twenty-four, Halyburton was one of the youngest officers onboard, and while he was proud to be part of a complex, powerful enterprise, he had quibbles with life at sea. The odor of jet fuel seemed always to permeate his clothes and hair, the omnipresent steam created an acrid smell, and the food was lousy. As he told Marty on tape, "The kitchen ran out of eggs, so breakfast doesn't hold much appeal to me. Lunches have been pretty bad, and dinners have been edible but nothing tremendous." Alcohol was forbidden, but Halyburton, like most officers, kept a fifth of gin and a fifth of Scotch in his footlocker.

He found refuge, if not exactly quiet, in his relatively spacious stateroom, where he lived with a lieutenant. (Most junior officers were quartered in crowded, six-person bunkrooms, but for some reason he was given better accommodations.) He kept a diary ("Feel the urge & need to write again. Imagine a poet-artist-RIO"), worked on some of his poetry, and read Ayn Rand and *Wind from the Carolinas*, a historical adventure novel. He recited inspirational passages from a small black vinyl book that his mother had given him. The typewritten letters did not always strike the narrow page evenly, but the slants and smudges brought a human touch to lofty truths.

> Tho all seems chaos now and
> Night prevails

Upon earth's wreck-strewn
Shores and blighted plains
Yet always after winter's
Cruel gales
Comes April with her
Iridescent rains . . .

These things shall pass, the
Wounding things of time,
And comfort to sustain is
Found in prayer.
Mankind is blest by lives
Pure and sublime,
The far reflection of the
Love we bear.

The hardest part of the cruise was the long separation from home. His daughter, Dabney, was born four weeks before he left, and Porter had seen her for only five days. Onboard, he heard plenty of stories about troubled marriages. The military life, with constant uprooting, long absences, and mediocre pay, was brutal on families, and Halyburton did not want to see his undermined.

He and Marty had met in college and had been married for only a year and a half. The pair cut a striking figure on the dance floor: Porter, with long, powerful strides, twirling his petite bride; Marty, her short blond hair flipped over the top of her head, smiling and laughing. On their first anniversary, Porter gave her a bottle of Estée Lauder perfume; she gave him an apron with his name on it. He read her passages from "The Rime of the Ancient Mariner." She called him "Julius," because when he walked out of the shower, his hair flopped over his forehead like Julius Caesar's.

She wasn't keen about his joining the Navy and flying jet fighters; surely there were safer ways to make a living. But she knew she couldn't talk him out of it and never really tried. Her perspective changed when Porter was in flight training in Key West and the spouses were allowed to visit a carrier. For the first time, Marty heard the roar of the engines and saw how the planes were "shot out of the catapult like a cannon." She was surrounded by the exhaust, the flames, the smoke; the din was constant and the crewmen ran about everywhere. She felt the quaking of the deck when the jets landed, and she saw the arrest wires stopping these great machines on a dime. She realized the entire operation was connected, communal — it was the most exciting thing she'd ever seen in her life. When they got home, she told Porter, "Now I understand why you want to do this."

6

"No Chutes Observed"

Halyburton considered himself fortunate in one respect. For most of his missions, he flew with Lieutenant Commander Stanley Olmstead, whose good looks, humble roots, and aeronautical savvy seemed lifted from a military recruitment catalogue. He was six-feet-one, with curly blond hair, blue eyes, and an easy smile. He was raised on a farm in Marshall, Oklahoma, and in the 1950s entered the Navy, where he excelled as a guided missile pilot and a test pilot and was also invited to join the Navy's elite Blue Angels. His dream, however, was to become an astronaut, which was not unrealistic, for many of the astronauts had been military test pilots. Vietnam would be his first time in combat, but he was eager to continue his education, enter the Apollo program, and fly to the moon.

Halyburton was initially wary of the thirty-one-year-old officer. Previously, Olmstead had piloted a single-seat F-8, but now he had to fly with Halyburton, a junior navigation officer — literally a backseat driver. Many lieutenant commanders would resent such an arrangement. A picture of the two men, wearing flight suits on the carrier's deck, reveals the contrast between them: Halyburton, his head tilted, his weight back, his arms pressed against his stomach, appears reticent, tentative; Olmstead, his chin out, his hands on his waist, exudes confidence and maturity.

Nevertheless, the pair worked well together. Olmstead sought advice from Halyburton, relied on his judgment, and made him feel like part of a team. Soon a friendship developed. For each flight, when one of them botched a radio transmission, he owed his partner a beer. They kept a running tally; Porter almost always owed Stan a drink or two.

The *Independence* was initially based off the coast of South Vietnam, and on July 1, 1965, Halyburton boarded the back seat of his F-4 Phantom for his first combat mission. Their task was to support the ground troops against the Communist insurgents, and Olmstead fired high-velocity Zuni rockets at the Viet Cong, destroying eleven huts. They had two more missions the following day. Within a week, the ship steamed north to launch flights over North Vietnam, and Halyburton flew almost daily for five and a half weeks. At first he found the bombing runs exciting. Many were at night, requiring them to use flares before launching the explosives. They attacked bridges, trucks, and boats, dropped napalm, fired sidewinder missiles, and took reconnaissance photographs. Halyburton liked the action, a pace he maintained onboard as the assistant weapons officer, which required him to be ondeck when ordnance was loaded. His unit packed more than a million pounds of bombs, missiles, and ammunition during his time at sea.

In August he had a two-week break in Japan. Marty, home with the baby, couldn't visit, and the respite only heightened Porter's homesickness. "I confess I broke down and cried a little bit the other night," he told her on tape. "I couldn't stand being in a beautiful place like this without you." He then returned to the flight line, and by the end of the month he was promoted from ensign to lieutenant j.g.; he would soon receive his first air medal. But by then doubts about the operation had set in. Mirroring Cherry's sentiments, he was frustrated by the petty targets, using million-dollar aircraft to drop bombs on the Viet Cong's water buffalo.

And the Vietnamese were crafty in their defenses. On a night raid, Halyburton would see the lights of a town in the distance, but the Vietnamese had early-warning radars and could hear the low-flying jet. As the F-4 approached, the town's lights shut off, the countryside went black, and the target disappeared. Trucks that carried supplies to the South were difficult to pinpoint as well. As Halyburton later discovered, the Vietnamese camouflaged the top of their vehicles with a green canopy. When a driver heard a jet, he drove to the side of the road, making the truck indistinguishable from the countryside. At night, the driver simply turned off his lights. Even when the Americans did hit a truck, they couldn't be certain whether it was military or civilian. Once Olmstead bombed a vehicle that Halyburton feared could have been an ambulance, but Olmstead believed he had no choice. They had been under antiaircraft fire, and he concluded that the civilian designation was meaningless when an entire country was shooting at you. "They're all Communists," he told Halyburton.

The bombing restraints were suppose to induce the enemy to the negotiating table, but in Halyburton's view the restrictions simply emboldened the Vietnamese, allowing them to move or safeguard vital assets, like oil tanks and factories. He was disappointed that they did not encounter any Soviet MiGs. Shooting down an enemy aircraft is every fighter crew's dream, and it also provides the greatest test for an RIO, who has to home in on a moving target. While Halyburton was still useful on bombing raids, with navigation, communication, and spotting landmarks, he felt less essential.

He was further distressed by the dishonesty of the flight reports. When the airmen returned from their missions, they were debriefed by an intelligence officer. One day Halyburton read a notice describing the different ways in which an aviator could describe an attack. But each outcome pertained to a successful hit, such as inflicting collateral damage, cratering a road, or destroying the target. Missing was not an option.

During one debriefing, Halyburton acknowledged that his bombs did not strike the target: "I think we missed."

"You're not supposed to say 'missed,'" the official responded.

"Well, you can say whatever you want. I'm telling you we missed."

Counting hits, real or imagined, was part of Defense Secretary Robert McNamara's statistical obsession, the belief that he could quantify his battlefield successes to justify the further escalation of force, even if the numbers were meaningless.

Nonetheless, Halyburton was proud of his flight performance — he would fly on seventy missions — and still believed that communism had to be stopped in Southeast Asia. But by the second week of October he was glad his tour would be over soon. He had one more stint on the line; then the *Independence* would be leaving on November 21, arriving home by Christmas. He could volunteer for another tour of combat, but he had already decided against it. He had done his duty, and the separation was becoming painful. He told Marty on a tape,

> Most of the time I don't feel sorry for myself . . . I feel like I'm doing something real worthwhile and not just for myself . . . You get a good feeling about it. But being away from you is just about more than I can take. I look at your pictures and read your letters and wonder what you are doing, what you are wearing, who you are with. I guess I'm really homesick now.

Marty was getting equally nervous. She watched the news three times a day and saw a report that an A-6 Intruder from the *Independence* was shot down, its crew presumably captured. "Oh, I just pray something can be done to bring an end to this fighting," she wrote to him.

Finally, on October 16, Halyburton received some good news. He was going on an "Alpha strike."

Alpha strikes carried high military importance — in this case, a bridge in the town of Thai Nguyen, a target that had previously been on the restricted list. The bridge, seventy-five miles north of Hanoi, was a major rail link between China and North Vietnam, for the railroad moved weapons that were ultimately brought into the war. The importance of the mission was made clear at the briefing, which was attended by Admiral Grant Sharp, head of the Pacific Command, whom Halyburton had never seen before. Halyburton's previous missions usually entailed four aircraft, but this one would involve thirty-five — a massive strike force. No one ever explained the change in strategy, and Halyburton didn't care. All along, he believed the Navy was dissipating its resources by trying to strike individual trucks or other incremental assets instead of attacking "hard targets" and inflicting real damage. Now was their chance.

On the morning of October 17, a Sunday, the strike force departed, with Halyburton's F-4 at the very end of the formation. The jets flew at three thousand feet above the water, refueled above the Gulf of Tonkin, and reached the coast forty miles north of Haiphong, avoiding its missile installations. Then they descended, dropping so low that, in the words of one pilot, "I could see individual grass spears." The low altitude enabled them to fly undetected by the radar at SAM installations, but it also put them in easy range of antiaircraft fire.

The size of the mission created another tradeoff. It added firepower but also integrated different types of jets — which specifically hurt Halyburton's fast F-4. That aircraft, by itself, would normally fly at about 550 knots, but because it flew this mission with the plodding A-4 Skyhawk, encumbered with heavy bombs, the F-4 had to travel at about 360 knots, which also decreased its ability to maneuver. Halyburton and Olmstead had bad luck as well with the position of their plane at the rear of the formation, increasing their exposure to enemy fire. The airmen were assigned not to hit the primary target, the bridge, but to destroy the surrounding antiaircraft sites used to shoot down planes.

The strike force, flying west for twenty minutes at treetop level, penetrated deep into North Vietnam — much farther than Halyburton had ever flown before. He knew how exposed they were and how they would have little means to resist or dodge enemy bullets or "flak," little gun bursts that would explode near a jet, spraying it with shrapnel. Fifty miles from the target and forty miles northeast of Hanoi, the jets flew over a hill and into a valley. They were suddenly about fifteen hundred feet aboveground, and Halyburton, looking straight ahead, saw karst ridges, the sheer limestone cliffs that were the marker for their next turn.

"Stan, we're coming up to our next point," Halyburton said while peering at his map. "We're within five miles of karst ridge."

"Roger."

Then Halyburton saw three puffs of black smoke as enemy gunmen, waiting near a railroad intersection, fired exploding projectiles at the American onslaught. The snipers may have had radar, hidden by the valley, that alerted them to the jets; the pilots later described the ammunition as the size of tennis balls. Then Halyburton felt a thump. The F-4 was still flying straight, but he knew they were hit. He tried to key his microphone, in his oxygen mask, to notify the other airmen, but when he pressed the button he realized that the mask had been blown off.

It was the first indication that he had seriously underestimated the impact of the shrapnel. He then leaned to the side and looked through a small tunnel into the cockpit. He saw Olmstead's head slumped over, his helmet off, papers blowing all over, and holes in the canopy. The cockpit had been hit from below, causing metal fragments to rip through the top. Halyburton realized that Stan was either dead or mortally wounded. He thought that if he had a control stick, he could turn the plane around and eject with some chance of being rescued. But there was no control instrument in the back seat of an F-4, and he knew the aircraft was doomed, though it continued to fly straight. Halyburton saw he was heading right for the karst ridge. In the seconds before impact, it occurred to him that he might be better off staying in the plane, that death might be better than ejecting into enemy hands. But he realized that wouldn't be right.

He was wearing flight gloves, but when he looked down he saw that a piece of metal was sticking out of his hand. He pulled it out. Then he reached up for his ejection handle but instead felt twisted metal — remarkably, his knee board, which had been strapped to his leg, had flown up and somehow impaled itself around the handle. The ridge was fast approaching, and he had only seconds to escape. He then reached between his legs for the backup ejection handle. If it didn't work, he would be dead in moments. He yanked it, the canopy blew off, and he shot out of the plane, his seat falling away. His parachute opened as he heard his plane, with Stan Olmstead, explode as it crashed into the karst.

Halyburton was saved but hardly safe. Descending, he heard bullets whip by. He hit the ground and tumbled into some scrub brush, suffering cuts and bruises. He pulled off his helmet and parachute harness and took from his pocket a green "survival radio," connected to a battery. Still dazed, he began yelling into the device. "Mayday, mayday! I've been shot down!" He heard nothing, however, not even static.

It was a warm, sunny day, and he began looking for a place to hide. He had few options. He took a step toward some bushes, but stopped after he saw a snake heading the same way. Trying to dis-

59

tance himself from a nearby village, he scrambled up a hill, where he could partially conceal himself. He wanted to keep running, but he was breathing heavily, his mouth was cottony, and a tall white American had no place to hide in a country of diminutive Asians. A rescue helicopter was his only hope, so he tried the radio again, screaming, "I'm on the ground! I'm in danger of being captured!"

But the radio's battery was dead, which infuriated Halyburton. The *Independence* didn't have enough radios for each aviator; he was simply given one before a flight, but he wasn't able to test it. Before a mission, the ship could not send out radio signals, which might reveal the carrier's position. This policy baffled Halyburton, as the *Independence* was already sending out scores of electronic signals and one more from a radio would hardly jeopardize it. Nevertheless, an airman never knew if his radio worked, and Halyburton's fears that he would have a defective one in a moment of crisis had just been confirmed.

He sat and listened for a rescue helicopter or a plane, but all he heard were villagers closing in on him. He had his .38 pistol, but it was a signaling device, shooting tracers, not a weapon with bullets. He decided to destroy his radio, lest the enemy recover and repair it, then use it to ambush American planes. Halyburton pulled out his knife, smashed the face of the radio, and cut the battery cord — a painful task. Even though the radio was useless, breaking it ensured that he would not be found and rescued.

Minutes later, he was surrounded by thirty or forty men, armed with rifles, machetes, or pitchforks. He stood up, raised his arms, and surrendered.

There was good reason why no rescue mission was sent for Halyburton. First, the Navy didn't know he'd been shot down; then it thought he was dead.

Shortly after his F-4 was hit, another Phantom was also struck, but the pilot, Tubby Johnson, radioed the strike to the rest of the mission. Moments later, Halyburton's jet exploded, leading most of the other airmen to believe that Johnson's plane had crashed. In fact, that jet, while losing an engine, was able to circle back and return to the *Independence*. However, none of the airmen saw Halyburton's parachute — most of them were too far ahead — so no one knew he lay in Vietnamese brush.

Only after all the crews had returned to the ship did they realize that Halyburton and Olmstead had been shot down. By then two other F-4s had suffered the same fate over the same valley on their way out of Vietnam. Incredibly, the mission took the same course leaving the country as entering it. While it was the most direct route, repeating it allowed the same gunners to shoot down two more jets. The four airmen from those jets were seen in their parachutes. Rescue helicopters searched in vain for them; they were listed as missing in action.

But Halyburton, obscured by the ridges and trailing the strike force, went unnoticed. As Lieutenant Al Carpenter, piloting an A-4 on the mission, wrote in his log: "Another not so good day for the *Independence*. In a big strike on a highway bridge at Thai Nguyen, we lost three F-4s . . . On the way in we ran into flak. Crossing a valley with a highway and a railroad in it, Cmd. Olmstead caught a good hit evidently and immediately ran into a karst hill and exploded. Very spectacular. No chutes observed."

There was one other way that Halyburton could have been found. He landed with a seat pan attached to his parachute harness; inside the pan was a radio that emitted a beeper signal, which would have been picked up by the airmen on the Alpha strike flying out of North Vietnam. But the signal was never detected because, one assumes, it was never sent. (More than thirty aircraft flying over the radio could not have missed it.) Halyburton thus landed with two radios, and both malfunctioned.

With no sighting of a parachute or evidence of a radio signal, the commander of Halyburton's squadron, Lewis S. Lamoreaux, concluded that he had been killed in action. The commander wrote in his final evaluation:

> LTJG Halyburton was an outstanding young officer, of great potential and value to the service. [He] flew more than 65 sorties against communist forces in Southeast Asia. For this he was awarded 6 Air Medals, the Navy Commendation Medal, and was recommended for the Distinguish Flying Cross. He lost his life when his aircraft was shot down by enemy ground fire while on a strike deep in enemy territory north of Hanoi on 17 October 1965.

The mission itself failed to destroy the bridge at Thai Nguyen. Al Carpenter wrote in his log: "BDA [bomb damage assessment] showed the bridge heavily damaged but still standing. No spans knocked down." Meanwhile, the antiaircraft site that Olmstead and Halyburton were to hit was instead targeted by Ralph Gaither, a young F-4 pilot who saw their jet crash. Gaither was also supposed to fire his rockets at an antiaircraft site near the bridge, but he couldn't find it. So he sought out the Olmstead-Halyburton target instead. When he drew near, however, the site was quiet — there was no gunfire. It was, Gaither concluded, nonoperational, a decoy, adding a painful coda to the mission: Halyburton and Olmstead were shot down trying to destroy a target that didn't exist. Gaither, for his part, was no luckier than Halyburton. He piloted one of the two other planes shot down on the Alpha strike, and he and his RIO, Lieutenant (j.g.) Rodney Knutson, were captured.

Despite the loss of three planes and six men (three captured, two killed, plus Halyburton) and the failure to knock out the bridge, the attack was heralded as a success. When the *Independence* returned to Norfolk on December 13, a front-page article in the *Virginian-Pilot* noted that the enemy had suffered mightily from 10,309 sorties that had dropped or fired more than nine million pounds of steel or explosives. Only one mission received specific praise — that of October 17, in which pilots "were credited with the first destruction of an active, mobile surface-to-air missile site in North Viet Nam."

The peasants surrounding Halyburton spoke no English, but he could usually figure out what they wanted. They stripped him of his flight vest, pistol, Winston cigarettes, anti-g suit, and boots. Tying his arms behind his back, they began marching him through low, rolling hills toward their village, almost two miles away. At the outset, he heard the fighter jets from his mission flying over the valley. The peasants pushed him face down on the side of the road and sat on him, but he could still hear the antiaircraft fire that would shoot two of the planes down. After the jets were gone, the group stood up and walked the rest of the way to the village.

Halyburton didn't know its name, but his treatment there was relatively benign. A large crowd met him, strained to get a better look, and followed him to a hut with mud walls. He figured he was the only white man who wasn't French that the villagers had ever seen, and he felt as though he were from another planet. With his hands still tied, he was a source of curiosity. As the villagers peered inside, he sat in a corner, had his hands untied — one was still bleeding from the cut in the plane — and was offered one of his own cigarettes. He pulled out a Zippo lighter, which was promptly confiscated. They feared he was going to ignite the thatched roof.

Desperate for water, Halyburton kept motioning that he needed to drink. The villagers initially brought some rice and soup, which Halyburton tasted out of respect; finally he was given water. The Vietnamese jammed inside the hut and poked through his belongings, which were fascinating but also dangerous. A farmer who picked up his pistol inadvertently fired it, sending a tracer through the hut. No one was hurt, though Halyburton feared that had anyone been shot, he would have been blamed. Such a mishap could have easily caused his execution. Meanwhile, his seat pan contained an inflatable eight-foot raft, and he was afraid the villagers toying with it would activate it. Anticipating pneumatic turmoil, Halyburton persuaded them to drop the device.

Shortly, some militiamen arrived and placed Halyburton in a Jeep, his gear in back. They drove through rugged country and eventually stopped at a stream. A soldier untied Halyburton's hands and gave him a canteen cup. But when he walked to the stream, he noticed that a militiaman stood with a camera, poised to click. Halyburton didn't want a photograph of him drinking water used for propaganda, but his thirst was overwhelming, so he began to scoop up the water — then stopped. The photographer took the picture, and Halyburton quickly drank before the cameraman could rewind. The ploy worked several times. Halyburton never knew how the pictures were used, but at least they did not show him drinking water.

The cat-and-mouse tactics soon ended. Halyburton was driven to a larger village, dropped off at a brick building, and told to sit at the end of a large table. Two guards stood at the door with Soviet AK-47s; others milled about the room and peered through two windows. An older man with a notebook and pen, a political cadre, sat to his left, and Halyburton knew this encounter would be rougher. His adversary couldn't speak English, but he had a book of translated English phrases. He copied down several questions and pushed the paper toward Halyburton, who read them to himself.

"What is your rank?"

"What kind of airplanes did you fly?"

"What was your target?"

Halyburton shook his head, and the interrogator slammed his fist on the table. He took back the paper, wrote out more questions, and slid it back. Again, the American ignored it. His defiance angered the guards, and one walked over and put the barrel of his rifle against his head. At this point, Halyburton didn't care if he was shot — he was not going to answer any questions. His attitude had less to do with loyalty to country than with crude calculations about his fate. He had heard horror stories about American POWs in South Vietnam: they had been executed and found with their heads severed, their genitals in their mouths. He had not heard how POWs in the North were being treated, but he thought he might prefer death.

He shoved the paper back. The gunman jammed the barrel against his head but didn't shoot. Halyburton soon noticed that the onlookers behind him were in the line of fire and realized that the soldier's job was to intimidate, to bluff. The gunman finally walked back to the door, though Halyburton felt the hatred in his eyes. The interrogator continued his work, but if anything, Halyburton felt even more belligerent. When the paper was again pushed his way, he shoved it back violently.

The move prompted the same gunman to return and put the rifle to Halyburton's head — but this time he moved a bullet into the chamber. Everyone scattered, and Halyburton thought, He's going to kill me. He heard the Vietnamese, civilians and soldiers, talking anxiously, apparently also not sure what to do. Halyburton thought of how much he loved his family, how much he would miss his wife and daughter, but he also believed that death right now might be better than life. He waited to die.

The gunman, however, didn't move. Others in the room calmed him down and finally removed him from the building.

Halyburton was unaware that any village capturing an American would be rewarded if it turned the prisoner in to the authorities. To him, it seemed a miracle that he wasn't dead, though his survival remained tenuous. When he was taken from the building, his arms still tied, the villagers gathered to throw rocks and clods of dirt and spit on him. He was put in the Jeep and spirited away.

It was dusk, and by the time the vehicle stopped, Halyburton was blindfolded. When the cloth was removed, he found himself in a room with two sawhorses, a board, a bed mat, and a bucket for a toilet. He was given some water. Everyone he saw had a weapon, so he assumed he was in the hands of the military. He figured if they had planned to kill him, he'd be dead by now; but he also reckoned that his treatment was going to get worse. Exhausted, he fell asleep but was awoken before dawn, again blindfolded, and taken away in the Jeep. This time, though, he could see beneath his mask, and he recognized Dave Wheat, an RIO who had flown on the same mission. He was thrilled to see another American.

The long journey continued until he saw tall, arched gates, high walls with barbed wire, and imposing buildings. This was no village. Then he saw something incongruous — raised flower beds in a courtyard. But there were no signs of welcome at the Hanoi Hilton.

Halyburton was locked in a room with a spigot. Craving water, he drank directly from the faucet — a huge mistake, he later realized; all the water at the prison was boiled before consumption. He felt bleary, but before he collapsed of exhaustion, he was taken into another room for questioning.

His first interrogator was Colonel Nam, the chain-smoking commander called Eagle who tried to convince the POWs that he was a MiG pilot. Halyburton sat on a low wooden stool and, as a sign of respect, was made to sit straight, his legs and arms uncrossed. When either was crossed, a guard would knock his arm. The early interrogations were perfunctory. Eagle asked about his plane, squadron, ship, and targets; the American stuck to name, rank, serial number, and date of birth.

Halyburton was then taken to Heartbreak Hotel, which some Americans compared to a frigid cement bunker. His cell had a cement bed with ankle stocks. A bare light bulb hung from the ceiling. Boards covered a window, and a small wastebucket stood next to a rathole. Gongs signaled the day's few activities. Martial music played from loudspeakers outside.

To wash up, Halyburton was taken to a cell with a spigot that dribbled only cold water. The prison still had Vietnamese inmates, and their civilian guards (as opposed to the military guards) would come into the Americans' corridor and urinate on the floor of the washroom. Halyburton, however, found some diversion when he entered the washroom for the first time, bent down to dump his bucket, and saw on the wall a circle with the words, "Smile, you're on *Candid Camera.*" That someone had retained a sense of humor lifted his morale.

He was even more pleased the first time he heard Americans whispering to one another. Above each cell door was a window covered by a wooden panel, and the inmates could stand on their beds, lift a panel, and talk. Of course, talking was forbidden, so they found other ways to communicate, such as whistling. When the prisoner on lookout wanted to indicate "all clear," he whistled a line from "Mary Had a Little Lamb." When a guard was coming, he whistled "Pop! Goes the Weasel."

Halyburton's flight suit was taken, along with his Rolex watch, a status symbol for aviators. He was given cotton pajamas, a tin cup, a straw mat, a thin cotton blanket, and a mosquito net. Two skimpy meals and three cigarettes a day provided moments of relief, though cigarettes would be withheld to punish his "bad attitude." He lost twenty-five pounds in a month and was down to about 150, but he still hoped he'd be home by Christmas. Like the other prisoners, he believed that the Vietnamese could not hold out against the bombing, that a settlement would be reached, and that the prisoners would be freed. To believe otherwise — to believe that the war would persist for six months or a year or even longer — would have been emotionally debilitating. No one anticipated such an outcome, or at least no one would publicly express it. For now, tedium and stagnation were often the enemy. When Halyburton wasn't being interrogated or confined in leg stocks, he walked back and forth across the cell — three steps and turn, three steps and turn — covering up to five miles a day. On Sunday the Americans had a "church call": everyone stood up, recited the Pledge of Allegiance, and said the Lord's Prayer.

His ability to resist the Vietnamese was reinforced by other Americans, notably James Stockdale. His rank (Navy commander), his intelligence (a graduate degree in international relations from Stanford), and his physical appearance (a weathered face and a shock of gray hair) all brought him respect, but his toughness inspired awe. When he ejected, he suffered a broken back and a fractured leg. He later pounded his face with a stool and against a wall until he was unfit to be photographed or filmed. When he entered Heartbreak, he feared he would die, so he told Halyburton, in an adjacent cell, everything that had happened to him in captivity. Halyburton, when released, was to speak to Stockdale's family.

Stockdale also urged Halyburton to remain strong during interrogations. "Stick to the code," he said. "Don't answer their questions, and don't ever think these guys are your friends."*

---

*Stockdale's heroics contributed to his winning a Congressional Medal of Honor and his selection as the running mate for Ross Perot in the 1992 presidential campaign.

But Halyburton knew the punishment the Vietnamese would inflict on an obstinate prisoner. When Rod Knutson was shot down on October 17, he did not surrender quietly, killing two riflemen in a shootout before he himself was grazed by a bullet. At Hoa Lo, he thrust the pen that he was supposed to use to sign a confession right through the paper, and he defied the ban on communications by yelling to other Americans in nearby cells. His truculence coincided with his captors' new latitude to brutalize prisoners. The guards locked him in ankle straps, bound his arms, denied him food and water, and punched him until his nose was shattered, his teeth broken, and his eyes swollen. Still defiant, he became the first American to suffer the "rope trick." He was pushed face-down on his bunk, his ankles put in stocks, his elbows bound tightly with manila hemp rope. The long end was then pulled up and attached to a hook in the ceiling. As the torturer, known as Pigeye, hoisted the prisoner, Knutson was lifted so he could not relieve any of his weight, causing his shoulders to feel as if they were being torn out of their sockets. He could barely breathe. Screaming and in tears, he agreed to talk.

The torture was both excruciating and diabolical, as it minimized obvious scars that could have been seen by outsiders. It also sent a powerful message to the other POWs: from now on, belligerence would come at a heavy price.

Halyburton chose to defy but not provoke — being respectful, he believed, carried no cost. Knutson was tough and even inspirational, but he could have avoided his beating. One of the leaders in the camp, Robby Risner, the Air Force lieutenant colonel, would later advise, "You catch more flies with sugar than you do with vinegar." But in the early stages of captivity, politeness did not satisfy Eagle, who began offering Halyburton the "better place–worse place" option. If he provided details about his missions, targets, and his own history, he would receive medical attention, better food, and more comfortable quarters. Halyburton didn't know what the "worse place" would be.

On the night of October 31, two weeks after his capture, Halyburton was blindfolded, handcuffed, placed in a truck, and taken to Cu Loc Prison. His small black cell with a bricked-up window reeked of wet cement, and he felt like Fortunato, the doomed character trapped in the wine cellar in Edgar Allan Poe's short story "The Cask of Amontillado." That night, his spirits declining, Halyburton prayed for strength. The next morning he awoke to a scratching sound from the mostly covered window. He looked out of a three-inch gap and saw a green leaf on a branch, which the wind had pushed through the shutter. Porter held the leaf and rubbed it between his fingers, grateful for evidence of life and, he felt, a sign that God could reveal Himself even in terrible conditions.

Over the next several days, a light bulb and bed board were brought in to him. More important, he heard tapping from the adjacent cell. It was Navy Commander Jerry Denton, a senior officer from the *Independence,* using a code common among the POWs.

Denton himself was a hard-line resister who had already been starved and placed in irons for his defiance. He told Halyburton that Risner was the senior commander at this prison and that Halyburton needed to contact him with information about himself and any other prisoners whom he had seen.

A few days later, Halyburton looked through the slats in his cell window and saw Risner in the yard, doing pushups. Guards stood nearby, but Halyburton tried talking to him through the window, identifying himself and naming other inmates. Halyburton didn't know if Risner heard him, but it felt good to be sharing "intelligence" and to be part of the resistance.

As the days passed, Halyburton needed to relieve his boredom and loneliness, so he and Denton decided to play chess. Halyburton had a piece of the coarse brown paper used for toilet paper. He folded it so that the paper had the right number of squares, then used a burning cigarette to mark the black squares. He saved pieces of bread and pinched them to resemble pawns, rooks, and the other pieces. Denton did the same thing so that each player had a complete chess board, and they signaled their moves by tapping on the wall, such as "queen's rook three to queen's rook five." Whenever a guard came they hid the board, but they did complete a game. As part of the crackdown, the Vietnamese began conducting room searches and punishing offenders. Halyburton couldn't risk getting caught with the game, so he ate the pieces. But the experience was a gratifying act of disobedience; while the Vietnamese could imprison his body, they could not destroy his imagination.

Halyburton underwent one or two interrogations a day, typically conducted by Rabbit, yielding the same "better place–worse place" impasse. Seven days after arriving at Cu Loc, he was moved to the cell known as "the auditorium"; it was smaller, darker, and more isolated than his previous space. The Vietnamese became more vengeful, screeching "Black criminal!" "Bad attitude!" and "Air pirate!" and forcing him to kneel on the cold concrete. The abuse mystified Halyburton, who assumed his captors would mistreat the senior commanders to send a message to the entire camp. Why would they single out a lowly lieutenant j.g.?

It was in the auditorium that Halyburton saw a beam of sunlight across the wall, which he marked with a piece of paper shaped like a cross. There was nothing unusual about the light, but like the leaf that poked inside his previous cell, Halyburton gave these signs divine meaning, drawing from them comfort and hope.

Denton had told him that if the Vietnamese had taken his picture, the CIA would know he was alive. So he assumed his family knew he had survived, but he despaired of his inability to contact anyone. In his final weeks on the *Independence,* his longing for Marty had become overpowering. Now, he couldn't reach her and didn't know how many days or weeks it would be until he'd see her again. To pass the time he continued to pray, but he also sang and whistled, cupping his hands to muffle the sound. He played "Am I Blue" and "Hit the Road, Jack," "Makin' Whoopee," and "Swing Low, Sweet Chariot" — pop songs, show tunes, and gospel. Like his imagination, his ear for music was beyond the reach of the enemy.

But Halyburton was now questioning his survival. He had not lost faith that the war would end soon; he just feared he wouldn't see it. He still had not disclosed any information beyond his name, rank, serial number, and date of birth; but he noticed that the interrogators weren't asking military questions anymore. The quizzes were increasingly used to lecture him on the centuries of exploitation suffered by the brave Vietnamese and the corruption of the American imperialists.

He stayed in the auditorium for ten days before being moved to the coal shed — a worse place, indeed. It was dark, cold, and filthy, and by now he was suffering from dysentery. The food had become steadily less palatable: he was receiving little more than rice infested with ants and soup with pig fat. He was surrounded by weeds, insects, geckos, and mosquitoes, he felt too weak to exercise, and he feared he was breaking down mentally and physically. He saw no more rays of light and heard no more scraping leaves.

Finally, he was taken to Fred Cherry's cell.

7

Strangers in the Cell

Y ou must take care of Cherry," the guard told Halyburton when they entered cell number 1. He left and locked the door.

It seemed, to Halyburton, an odd punishment. He was now in the same building, the Office, where he had been initially imprisoned at the Zoo. The cell, ten feet by twelve, had teak boards for beds and hooks for mosquito nets. A light bulb hung from the ceiling, and an adjacent cell allowed inmates to tap messages. His cell also had a roommate, Halyburton's first since his imprisonment. Having spent three weeks in dark, isolated rooms, he suddenly had contact with another human being. Close to a breakdown, desperate to talk to anyone, he was grateful to see Fred Cherry. But he was soon puzzled.

Cherry was clearly in bad shape. His left foot in a cast, his left arm in a sling, he had not washed in more than two weeks, and he had a hangdog, melancholy look on his face. Halyburton assumed that Cherry, wearing a standard olive prison uniform, was an American POW. But when Cherry said he piloted an F-105 as a major in the Air Force, Halyburton was doubtful. He had never known a black pilot — he wasn't even sure that blacks had the depth perception to hold such a job — and he certainly had never known a black major, which was two ranks higher than his own. Halyburton had never met a black who had outranked him — most African Americans he knew were laborers or domestics. He wasn't sure what was fact and what was fiction about Fred Cherry.

Cherry had even graver doubts about Halyburton. He didn't believe that he was a Navy lieutenant. He figured the Vietnamese would try anything to make him talk, so he thought Halyburton was a Frenchman spying for his captors. Halyburton didn't say or do anything to suggest such an identity; but Cherry's experience at survival schools had prepared him for Vietnamese duplicity, and he knew about the history of French colonialism in Indochina. It made sense to him that his handsome roommate would be a French spook, whose southern accent was one more clever ploy to throw him off the scent.

On their first night together, Halyburton hung his mosquito net and tried to make conversation, asking Cherry questions: When was he shot down? Where had his flight originated? Where was he from? These were many of the same questions the Vietnamese had been asking, confirming Cherry's suspicions. He either brushed them aside or told lies.

"I flew out of South Vietnam," he said.

"I was shot down a couple weeks ago," he reported.

He saw no reason to fib about his injuries, however. He said he had a dislocated shoulder, a broken ankle, and a cracked wrist.

Halyburton was disappointed at his frosty greeting and remained skeptical of his credentials, but he was still glad to have a cellmate. The gong sounded, signaling bedtime.

"Goodnight," he told Cherry.

Cherry played along. "Goodnight."

When Halyburton woke up the following morning, shafts of light penetrated a three-inch gap in the bricked-up window, and he got a better view of Cherry. He noticed that his forehead had a cut and that the ankle cast was so loose that it didn't appear to do much good. Cherry described how he injured his shoulder during his ejection, and Halyburton realized that he couldn't move that arm. It just hung limp in the sling.

"Does it hurt?" he asked.

Cherry shrugged. "If I move it," he said.

Halyburton was relieved that he was no longer isolated from other cells. He began tapping on the wall, and a response came from Rod Knutson, the Navy RIO who had been with him in Heartbreak. Halyburton described his grim journey through the Zoo and explained that he was now with Fred Cherry. Knutson knew of Cherry; he thought the Vietnamese may have intentionally placed a "southern gentleman" with a black officer to cause each man additional stress. This worried him, though he didn't disclose his concerns to Halyburton.

Halyburton himself had already recognized the enemy's motives: if the Vietnamese couldn't induce him to cooperate through harassment, abuse, and isolation, they would lock him up with a disabled black man and force him to care for that person or watch him suffer and possibly die. Either outcome, they believed, would be torment. The prison officials knew that Halyburton was from the South and that his older black cellmate would have military seniority over him — a reversal of authority for any white Southerner. They had also primed Cherry to distrust white Americans. During his interrogations, they repeatedly told him that whites were racist, that they were colonizers, and that he had far more in common with the colored people of Asia. They also invoked Malcolm X, the militant black leader who had been an early critic of the war. In 1964 he said his government was the most "hypocritical since the world began" because it "was supposed to be a democracy . . . but they want to draft you . . . and send you to Saigon to fight for them," while blacks still had to worry about getting "a right to register and vote without being murdered."

Had Cherry accepted these sentiments, the racial divide between him and Halyburton would have been insurmountable. But Cherry was less concerned about his government than about this stranger in his cell. On their first full day, Halyburton wasn't acting like a racist or a French spy. His communication through the wall with other Americans perplexed him — how could a spy do that? — but Cherry still didn't believe that he was a Navy lieutenant.

His skepticism lessened when Halyburton talked about the most obvious thing they had in common: they had both been shot down over North Vietnam. As it happened, the shootdowns occurred five days apart in roughly the same area, about forty miles north or northeast of Hanoi. "Maybe the same gunner got both of us?" Halyburton said, causing Cherry to chuckle. Each man described his encounters with the peasants and militiamen in the countryside, and the similarity of their experiences reassured both men about the identity of the other.

Their suspicions were further dampened when they began naming the other American captives they knew. Halyburton shared these names with Knutson, who in turn named the captives he knew. This process confirmed a basic military tenet: find out who's been injured or captured so you leave no man behind. When the discussion drifted away from military matters, Halyburton mentioned several counties in his home state of North Carolina — all familiar names to Cherry, whose father had been born in North Carolina and who grew up in Virginia.

Their forced intimacy evolved into an atmosphere of tolerance. They shared a rusty wastebucket, for example. When one was using it, the other stood in the opposite corner, his eyes averted. At least Halyburton's dysentery soon passed, making a terrible experience a bit less awful.

Their conversation became more relaxed, and Halyburton made his first offer of assistance.

"When did they let you bathe, wash up?" he asked.

"Bathed! I haven't bathed since I've been in this camp."

"Well, you should go every three or four days. I'll ask the guard."

In the early days, Halyburton taught Cherry the tap code, a means of communication that integrated him into the society of captives while also eliminating any distrust between the two men. Halyburton had discovered the code in Heartbreak. Visiting the washroom, he saw that someone had inscribed on the wall a matrix with five rows across and five rows down, a different letter in each position, and realized it was a tap code. Most of the inmates at Heartbreak ignored it, for they could usually talk to one another, but Halyburton worried that the time would come when that wasn't possible. He wrote the matrix on the liner of a cigarette pack and became one of its most avid practitioners. Learning the code gave him a chance to exercise his mind, and it was an act of defiance against his captors — in his words, "part of the tradecraft of being a prisoner."

To use the code, a person identified a letter by tapping out two numbers, the first giving the horizontal row number, the second, the vertical column number. A favorite message was 2-2 (G); 1-2 (B); and 4-5 (U) — GBU, an abbreviation for God Bless You, which became the universal signoff among the Americans.

The matrix had been carved into the wall by Air Force Captain Carlyle Harris, who understood that the Americans could orga-

	1	2	3	4	5
1	a	b	c	d	e
2	f	g	h	i	j
3	l	m	n	o	p
4	q	r	s	t	u
5	v	w	x	y	z

nize and resist only if they could communicate. When he was in survival training, an instructor had showed him the code during a coffee break, and this casual reference proved to be the lifeblood of the POWs in Vietnam. Whenever a newcomer arrived, the other prisoners' first responsibility was to teach him the code. Sometimes the Morse Code was used to explain it. Sometimes notes were slipped into rice bowls or wastebuckets. An American on the floor in one of Hoa Lo's torture rooms could find the matrix carved underneath a table with the words: "All prisoners learn this code."

Tapping became so routine, so pervasive, that many of the POWs could click as fast as they could think. Some spent years tapping to each other without ever seeing their partner; others developed playful shortcuts. Before two POWs went to sleep, one would tap "GN" (goodnight) and the other, "ST" (sleep tight), which would prompt the first person to respond "DLTBBB" (don't let the bedbugs bite). With only twenty-five slots, the matrix had no "k," forcing tappers to substitute "c." Thus, a popular transmission was "Joan Baez succs," sent after the Vietnamese played a tape of the American antiwar activist through the cells' loudspeakers.

The code was later adapted to different settings. A POW walking through a courtyard could use hand gestures like a third base coach to spell out words — scratching his head meant row one; touching his shoulder, column three — and other prisoners could watch from their cells. What the enemy may have thought was a nervous tic was actually a communication channel and a source of unity among the Americans.

To teach Cherry the code, Halyburton used cigarette ash to draw the matrix on toilet paper. By now he believed that Cherry was indeed an Air Force pilot and recognized that he had to learn the code. But Cherry saw a much deeper significance in his tutorial. Having watched and listened to Halyburton's "conversations" with Knutson, he knew they used the code to transfer classified information, so Halyburton would teach it to him only if he trusted him explicitly. Moreover, Cherry had not spoken to any Americans since his capture, so learning the code was like a child's learning to talk — haltingly, he got his voice.

He made mistakes while practicing, confusing the horizontal numbers with the vertical.

"What the hell is that?" Halyburton asked in mock anger after Cherry tapped a message.

"That's how you taught me," Cherry said.

"Well, you learned it outta phase." He could needle Cherry about the error without offending him.

Cherry could now be assimilated into the rest of the prison and take an active part in resisting the enemy. In his eyes, Halyburton's efforts were the act of a patriot. No spy would lend that type of assistance.

Even in a two-person cell, a chain of command was essential.

In every military organization, a command structure ensures accountability and discipline: you follow your senior officer or you're gone. In a POW camp, the structure provides a cohesive front against the enemy and reassures the prisoner of his military status. The Code of Conduct specifies that a POW must adhere to senior authority, which increases the chance of survival for himself and his cohorts.

In Vietnam, the POWs decided one chain of command should exist for all the services, not separate lines for each branch. Seniority was determined by rank at the time of shootdown, though promotions while in captivity were also considered. Sometimes the Vietnamese tried to undermine the command structure by segregating the senior officers or identifying a junior officer as "room responsible," but prisoners found other ways to communicate to preserve the chain. As long as communication was possible, each cell, each cell block, and each prison had an inviolable chain of command.

Once Halyburton accepted that Cherry was a major, that made him the cell's junior officer by two ranks. He privately noted the oddness of that arrangement, but Cherry's accommodating style — he didn't boss or bully or even give orders — eliminated any real concern. Cherry led by example and provided guidance instead. During a joint interrogation at Christmastime, for example, the Vietnamese offered the men candy and extra cigarettes. Halyburton looked at Cherry, who nodded, and Halyburton knew he could accept. It was a small gesture, perhaps imperceptible to the enemy, but still a significant departure from Halyburton's past.

Halyburton grew up in Davidson, North Carolina, a small college town where front doors were never locked and a peeping Tom signaled a virtual crime wave. Only twenty miles from Charlotte — the lousy roads made it a long twenty miles — Davidson considered its isolation part of its charm. It took snobbish pride in being an intellectual redoubt in the Carolina Piedmont, a hamlet of inquiry amid large oak trees, blooming azaleas, and crickets. Poetry readings, foreign films, international musicians, and renowned ministers were all part of campus life, and during Porter's teenage years, Robert Frost, Ogden Nash, Carl Sandburg, Isaac Stern, Louis Armstrong, and Dorothy Thompson passed through town.

The influence of the college probably made Davidson more racially tolerant than most of North Carolina, but it still resisted desegregation and civil rights. To Porter, overt signs of racism abounded. His friends called Brazilian nuts "nigger toes," and adults used racial epithets. He saw a merchant use a cigarette to burn a black youth he suspected of mischief. On Halloween, he and his friends were allowed to visit only those houses owned by whites. The black barber would cut the hair only of white patrons; his customers would have gone elsewhere if he had used his scissors on the head of a black man. In 1955, when Porter was fourteen, the town newspaper ran a front-page editorial outlining its opposition to integration, citing the Negroes' alleged stunted intellectual development, their propensity for crime, and their loose morals. "Many white persons believe that morals among their own race are lax enough without exposing their children to an even more primitive view of sex habits," the *Gazette* said.

While such views may have reflected the attitudes of many Davidsonians, they were not held by Porter's family, whose decency and decorum produced a different kind of racism — less hateful but in some ways more insidious.

Porter lived with his mother and grandparents. His parents had divorced when he was very young, and he neither saw nor spoke to his father. His principal role model (and his namesake) was his grandfather William Porter, a short, stout figure whose progressive ideas sometimes caused him trouble. In 1921 Davidson College hired him as a biology professor, but after he taught evolution, he was reassigned to the geology department. Passionate about self-improvement, he read the Bible in Urdu, worked complicated crossword puzzles called anacrostics, and watched television without the sound. "If I ever lose my hearing," he explained, "I'll be able to read lips."

William Porter employed a black domestic for many years and treated her kindly, paying for her and her sister to attend beauty and nursing school. Each Christmas their elderly father would come to the house for his gift; one year when he came to the back porch, Porter said, "You can come to the front door and get it." He would never have felt obliged to give school funds or a holiday gift to poor whites, who, he believed, could earn their money through hard work. Blacks were different. They needed help, and it was the responsibility of good whites to provide it.

To young Porter, his grandfather, as well as the entire community, reinforced the idea that blacks were limited. Whites held jobs that required creativity and intelligence — professors, writers, doctors. Blacks worked as janitors, postal clerks, cooks, garbage collectors, and yardmen. Their labor was indispensable to the functioning of any town, so they should be treated respectfully — but not equally. In his own house, blacks could enter through the front door, but they were not allowed to use the same bathrooms.

Halyburton, through his segregated high school, college, and military training, paid little attention to the civil rights movement or to its increasingly strident protests. The marches, the sit-ins, the freedom rides: they happened in the South but were not of his world. In his poems and other writing, Halyburton showed his sensitivity to many things — his friends, his family, the seasons, the environment, his own emotions. But on race, he was comfortable with the status quo. His attitudes were reaffirmed on the *Independence,* where white officers flew magnificent jets while enlisted blacks washed clothes, fixed meals, and tightened screws. Halyburton did not endorse the violent bigotry of cross burning and bludgeoning but embraced the soft prejudice of paternalism. He was proud of what he considered his grandfather's racial beneficence, yet ignorant of his own condescension.

Until Halyburton met Fred Cherry, he had never deferred to the judgment of a black man. Even more unusual was his handling of all the domestic chores.

Because he could use both hands, he could help Cherry as an orderly would help a patient. The guards had given them a whiskbroom, so each day Halyburton swept the floor. He picked up the bowls of food left outside the door. He emptied their wastebucket. He hung up Cherry's mosquito net at night and lit his cigarettes. The men were permitted to shave once a week, using cold water and old blades, but Halyburton began to demand from the turnkey that Cherry be allowed to shower.

Halyburton was mindful of their reversed roles — a white man doing chores for a black man — but he didn't care. The tasks gave him something to do and relieved the boredom. His efforts, however, made a deeper impression on Cherry, whose own experience had not anticipated a living arrangement with Halyburton.

Born in 1928, Fred Cherry was raised on a small farm outside Suffolk, Virginia, a region known for a massive swamp that covered 400,000 acres across the coastal plain of Virginia and North Carolina. It was known as the Great Dismal, a name that evoked the harsh conditions of the region's black families and the sense that they were at the mercy of forces beyond their control: segregation, disenfranchisement, unemployment, poverty, disease. Most black men had to rely on their backs to squeak out a living. They spent their days planting, weeding, and picking corn, potatoes, or cotton. In 1930 only nineteen percent of rural black families in the Lower Tidewater owned their own land; the rest were tenants or laborers. Those who didn't farm worked in peanut factories, iron foundries, lumber mills, or a naval shipyard. Few if any went to college, and many did not finish high school because classes overlapped the harvest season.

The youngest of eight children, Fred was a small, wiry boy whose penchant for jumping, running, and scurrying about led others to say he had a lot of pep — so they called him Pepper. Like many rural blacks in the South, the Cherrys adhered to the philosophy of Booker T. Washington, the former Virginia slave whose response to white supremacy had been interracial cooperation, the encouragement of thrift and business among blacks, and the acquisition of land. In 1881 Washington founded the Tuskegee Institute, its mission to train the "head, heart, and hand" of students who would then elevate the race culturally, socially, and economically. Progress would come not from dissent or confrontation but from self-improvement and accommodation. There were few alternatives. In the early decades of the twentieth century, the impoverished, agrarian South, with its history of black servitude and subjugation, was no arena in which to demand meaningful civil rights. The status quo would not survive; but long after the civil rights movement had roiled the South, the ethos of that earlier milieu — achievement through hard work, compromise, and accommodation — would continue to define Fred Cherry.

His mother, Leolia, rarely talked about race, but she did warn him about one of the great taboos of the South. "If you ever say the wrong thing to a white girl," she said, "you're dead." Fred had plenty of reminders of "blacks' place" in society. White children rode buses to school; Fred and his friends walked two and a half miles to theirs. Fred lived near a white family, the Gregorys, and whenever his mother sent him to borrow sugar or butter, she reminded him to address the adults by their last names preceded by "Mister" or "Miss." But Fred noticed that the Gregory children called his parents "Leolia" and "John."

Other examples of racism were not so subtle. Fred was once riding his bicycle with his older brother James along a two-lane country road. A car with two white teenagers pulled alongside; one reached out and knocked James on the back of the head, spilling him to the ground. The teens drove away, laughing.

Both James and Fred were furious, but their mother urged them to forget the incident. "James is all right," she said. "He's not hurt, so don't worry about it. You don't know who it was, so you can't mention it to their parents." Fred wasn't happy about it, but he realized his mother was probably correct — blind vengeance would be counterproductive. He also learned something else: to succeed in a white world, he would have to be a little bit tougher than everyone else, maybe a lot tougher. He remembered something his father had always told him: "Some things you can change and some things you just got to put up with. It's up to you to figure out which is which."

Racism didn't change after Fred became an Air Force officer. By leaving his insular black farming community, he encountered a more overt bigotry.

After serving in the Korean War, Cherry returned to the United States and, assigned to a new base, drove his family across the country. Though he often wore his well-pressed uniform off the base — dress blues or khakis, a silver bar on his collar — restaurants and motels routinely denied him service. But he never complained. And when traveling with white companions, he would decline their invitations to diners that he knew did not accept blacks, for he did not want to make a scene. He knew that confrontation of any kind could jeopardize him.

Housing was another problem. In 1957, when he was assigned to the Dover Air Base in Dover, Delaware, the town's black residential section had no vacancies, so his wife, Shirley, and their two sons initially lived with his family in Suffolk, Virginia. Cherry lived on the base until a "colored apartment" became available. Dover's segregation was even more unseemly in its public hospital, where Shirley gave birth to their first daughter, Debbie. The segregated wing and nursery were anticipated. Not expected was the unwillingness of the white nurses to bathe or change the baby; they just gave her to Shirley wet or dirty. Shirley complained to Fred, who demanded that his child be treated properly. When he returned the next day and saw that the neglect had continued, he finally lost his temper, cursing the nurses and telling his wife,

"We're leaving!" He grabbed the baby, and as the three of them headed for the door, a nurse yelled, "You have to check her out." Fred shouted back, "The hell with you — we're gone!"

In the military, he often had to overcome others' perception of inferiority, and his cause was not helped by the dominance of white Southerners in the armed services' leadership. Cherry was the first black cadet sent for basic training to Malden, Missouri, whose drab air base sat amid cotton fields in the Missouri Bootheel. His flight instructors, hired as civilian contractors, refused to teach him, forcing the commander to offer a promotion to any instructor who would.

Cherry's peers didn't treat him any better. When his flight class walked across the tarmac, the four white students walked in one line and Cherry walked by himself. But it didn't bother the new recruit; he was accustomed to segregation and only cared about being a combat pilot, not making friends. Besides, his piloting skills won the others' respect: after he was the first member of his class to fly solo, the other students walked with him.

Later, discrimination robbed him of one of the most exciting assignments of the Cold War.

In 1955 Cherry was stationed on Malmstrom Air Force Base in Great Falls, Montana, where he was one of several pilots recruited by the CIA for a highly classified mission: flying the U-2, a high-altitude spy plane with sophisticated avionics and reconnaissance equipment, over the Soviet Union to photograph its military capacity.

Cherry needed little convincing to participate in something bold, secretive, and dangerous.* The highest he had ever flown was fifty-four thousand feet, whereas the U-2 would ascend to

---

*The risks of the U-2 were confirmed in 1960, when Gary Powers was shot down over enemy air space.

ninety thousand feet. With several other pilots in his squadron, he was cleared for the program and awaited orders to leave the base. One day, however, he realized something had gone wrong when the other pilots were packing up without him. He called his CIA contact, who promised to find out what was going on.

Calling back, he gave Cherry the news. "I'm sorry, but your folder has been removed from the rest," he said. "We can't keep you in the program." The CIA had nothing to do with his removal, the official said, but an Air Force lieutenant colonel, who had to approve the transfer, had pulled Cherry's folder.

He didn't mention race, but he didn't have to — it was understood. Each personnel file includes a mug shot, so the lieutenant colonel, believing that a black officer was not fit for such a sensitive position, squashed the transfer. There was nothing Cherry could do, no one to hear his appeal. His own commander had not even been briefed on the program. So his career as a spy pilot ended before it began, though he occasionally wondered what the earth looked like from ninety thousand feet.

Cherry had another reason to distrust Halyburton. Porter was with the Navy.

His bias against the Navy was partly based on the rivalry between the Navy and the Air Force. From the time of the War of Independence, the Navy considered itself the service of history and tradition and viewed the Air Force, established in 1947 as an offshoot of the Army, as high-tech parvenus. According to the Navy, its airmen possessed superior skills to take off and land on a ship, but they also had many other duties onboard (like tracking ordnance) while living in cramped quarters. Air Force pilots could focus exclusively on flying and lived on roomy bases. While limited ship space forced the Navy fliers to economize on equipment and material, the Air Force could splurge on extra radios and other accessories. Halyburton was surprised to learn that Air Force pilots could jettison empty external fuel tanks for greater mobility; Navy pilots, lacking extra tanks, could not. In their dark blue with gold trim, the Navy airmen viewed Air Force apparel — a lighter blue with gray trim — as utilitarian and called the pilots "bus drivers."

The Air Force was no less derisive of the Navy, which it considered stodgy and aristocratic. As the newest military branch, the Air Force was devoted exclusively to air power, and it believed it operated on the technological frontier. It attracted large numbers of educated engineers, scientists, and mathematicians, who made it the military's most progressive service. Not surprisingly, the Air Force integrated more rapidly than any other branch. The Navy was the slowest.

Fred Cherry knew about the Navy's dismal record in race relations firsthand, and he wanted to tell Halyburton about his experience. In 1949, he explained, he attended Virginia Union College, one year after President Harry Truman had signed the executive order desegregating the military. "In my second year at school," he said, "I heard that if you qualified in all respects, you could go from civilian life straight into aviation cadet training. So I went to see this Navy recruiter in Portsmouth. I didn't know the Navy didn't have any black aviators. The recruiter told me to fill out this application for enlisted service. I said, 'No. I want to be a pilot.' Then he told me that the individual I would have to talk to was not in the office, and I could stop in some other day."

He went back three more times, he told Halyburton, and each time was told the commander was out. The fourth time, Cherry said, "I saw this door creeping closed. I knew he was there and had been there every time before. I just sort of exploded. I kicked the door open. He thought I was coming across the desk. I said a few choice words to him. They were rather obscene. Then I told him I didn't want any part of his Navy."

From that day on, Cherry believed the Navy was "a bastion of racism," though he hadn't met another Navy man until Halyburton walked into his cell. Now he began to revise his judgment and recognize that progress had been made. Halyburton was embarrassed and tried to apologize for the Navy. "I can't believe the recruiter would treat you that way," he said. He knew that Cherry still resented his service but was glad he didn't bear that grudge against him.

Common ground was easy to find. Baseball, for example. The prisoners weren't fans of the major leagues, but each had played as a kid, Cherry in a cow pasture and Halyburton on a sandlot. Each knew the social order of the South and the very different circumstances of his upbringing. But baseball was safe, and their stories — of running, catching, and hitting a ball in the sunshine of their youth — were easy to embrace. They could also share war stories, tales that offered action and suspense while not forcing either man to disclose much of himself personally. But Cherry was feeling more comfortable with his cellmate, so when he described a daring airborne rescue effort, he ended the story with a painful twist.

In the waning days of the Korean War, Cherry and another pilot, both flying F-84Gs, were attempting to land on a base at Teague, but the other jet's nose gear didn't lock. "I called the tower and pulled up beside him," Cherry told Halyburton. "They cleared the traffic pattern but he was getting low on fuel. He tried his emergency gear lock, but that didn't work. The gear was still loose, so I told him to give me a ten-degree bank and hold steady. I was going to see if I could knock it down, and he said, 'Okay.'"

Cherry acknowledged that the idea was risky — others might call it outrageous. He wanted to use his wing tip to nudge the landing gear into a locked position, but beneath the wing of each plane were the fuel tanks. The quirky angle would also force

Cherry to fly "uncoordinated" — in effect, without the usual navigational tools that ensured precise flying.

Nonetheless, at three thousand feet the impaired F-84G went into a slight bank, and Cherry slid right beneath him, steadied his aircraft, and kissed the landing gear locked. "It was dangerous," Cherry said, "because if his fuel tank bumped my plane, it would rupture and catch fire. If I touched him anyplace other than where I touched him, it would have been disastrous."

Halyburton asked where he'd learned such a move.

"I'd never heard of anyone doing that before," Cherry said, "but I felt if he could hold the aircraft steady, I could do it."

That wasn't the end of the story. Back on the base, Cherry went to the officers' club for a drink, eager to share his daring exploit. But he was never given a chance. Instead, he sat alone while a group of white officers fraternized among themselves. "They never said a thing," Cherry said.

But it was also the mindset of the military; it had been integrated in 1948 but not purged of its entrenched bigotry. While blacks had fought with distinction in American wars since the Revolution, they had often been ignored in official accounts or simply denigrated as unfit. In an influential 1925 report, the Army War College drew on racist anthropological studies to determine that blacks, with their "smaller cranium, lighter brain, [and] cowardly and immoral character," were lower on the evolutionary scale than whites, and they should be relegated to "special status" in the Army.

Integrated troops posed a threat by leaving open the possibility that a black officer would command white enlisted men. Even in the face of urgent manpower needs during World War II, maintaining the racial hierarchy was imperative — a point made by Secretary of War Henry L. Stimson in his private diary: "Leadership is not embedded in the Negro race . . . Colored troops do very well under white officers but every time we try to lift them a little beyond what they can do, disaster and confusion follow." President Truman's decision to integrate the armed services defied the country's military leaders, including George Marshall and Dwight Eisenhower, who had opposed such a move after World War II.

Cherry's slight at the officers' club reminded him of his second-class status. While he received glowing press coverage for his heroics, the incident was a sad postscript that he had carried with him for many years but had told to few people. In fact, he rarely volunteered stories about any of the discrimination he experienced, believing that complaints were useless and preferring the stoicism urged by his parents.

But he saw something different in Halyburton — a white man who would understand. Halyburton's relative youth seemed to give him an open-mindedness and sensitivity that he had not often seen in white Southerners. He appeared to know the difference between right and wrong. "He was more genuine," Cherry recalled.

Of course, in an earlier time Halyburton could have been one of the white officers who ignored Cherry, but he was now forced to reconsider his racial assumptions. He had been impressed by Cherry's travels — Japan, Thailand, Germany — and by his breadth of experience and piloting skills, his rescue of the F-84G being the most conspicuous example. Fred Cherry was unlike any black man he ever knew or heard of, yet his snub at the officers' club sent a powerful message: Cherry was in the military, but that didn't mean the military was truly integrated.

Their common problems drew them closer. To begin with, they were cold. While the temperature rarely dropped below freezing, the high humidity created a penetrating chill. Cherry and Halyburton each had one thin blanket, which was too short to cover them while sleeping unless they lay in a fetal position. They discussed this problem at length, analyzing the exact position that would maximize the blanket's coverage of the body. Compounding the problem was the concrete floor, which intensified the cold against their bare feet. They noticed that other prisoners were wearing shower shoes, and they wondered why they didn't have any. Their anger at the enemy spilled over into envy of the other POWs — and it was one more issue that linked them.

Their contempt for the Vietnamese sometimes fueled resistance. When a guard entered the cell, for example, he required them to show their subservience by standing and bowing their heads. One snarky guard was nicknamed "McGoo"; his squinty eyes evoked the visually challenged cartoon character. While most guards accepted a slight nod of the head, McGoo made it clear that that was not enough.

"Bow!" he yelled.

The Americans stood but didn't bow.

"Bow!" he repeated.

They nodded slightly, so McGoo walked over and slapped the heads of both men.

"Bow!" he said.

Their heads bobbed, but they would not bow.

There were more commands and more slaps, but the prisoners never complied, and the frustrated McGoo eventually left. Individually, Cherry and Halyburton had balked before, but together they pushed their resistance further.

To demoralize the POWs, the prison officials installed the same kind of loudspeakers in their cells that the government used to disseminate propaganda in the countryside. There, multikilowatt boxes were set up in hamlets that didn't even have water or sewage — indoctrination took precedence. Inside the prison cells, the green boxes typically delivered two broadcasts a day from the

"Voice of Vietnam," one intended for a general English-speaking audience and one for the GIs in the South. For two hours a day, the audio barrage was made worse by its blaring volume, what one POW described as "two decibels above the threshold of pain." The content included tributes to Ho Chi Minh on his birthday and long tutorials on Vietnam's history, including one frequent segment on its victory over the French at Dien Bien Phu. Some broadcasts were designed to mock the Americans. At the Zoo, one played a violin rendition of "Smoke Gets in Your Eyes," a reminder of the inmates' fiery ejections.

Halyburton and Cherry tried to glean some facts from the distortions. The broadcasts often specified how many U.S. aircraft had been shot down, but the two Americans, using information from other prisoners, estimated that the true number was probably one-tenth of the announced figure. The men laughed at some of the more outlandish statements and derided the mispronunciations — the city of Tucson, Arizona, for example, was pronounced "Tuck Sun."

But other news was more difficult to discount. Finding the names in *Stars and Stripes*, an announcer would recite a list of American casualties, described as "comrades who gave their lives in a needless, illegal war," with a violin dirge in the background. Once, the announcer named a Marine drill sergeant who had instructed Halyburton as an air cadet in Pensacola, Florida. The news of his death, exploited by the enemy, was disturbing, but over the years Halyburton and Cherry were even more enraged by the broadcast of antiwar statements from Joan Baez, Stokely Carmichael, and Ramsey Clark.

One time, Halyburton heard a recording of U.S. Senator Ernest Gruening of Alaska, who said that Americans who had been killed in Vietnam "have not died for their country . . . but have been mistakenly sacrificed as part of an inherited folly." Like virtually all of the prisoners, Halyburton despised the protesters — the actors, the college students, the hippies — but he held the politicians in particular contempt. He believed they provided comfort to the enemy, weakened America's will to win the war, and prolonged their imprisonment.

The Vietnamese reveled in these taped denunciations, which provided the very propaganda that the POWs resisted under mistreatment or torture. Once, Cherry was in leg stocks, tied to a bed, when he heard Jane Fonda accuse the POWs of cowardice for bombing children at night. Outraged, he "tried to tear his irons from the wall."

In the early days of December, the cold remained their worst problem, but their hunger was also acute. Food was precious, and each man's response to the meager offerings became an important — though unspoken — subplot in their relationship.

Twice a day, guards left bowls of food, usually soup, greens, and bread, outside their door. Halyburton would bring them in and serve Cherry, who would devour his meal with his one good arm. One day Halyburton noticed that one bowl had considerably more food than the other. Though Cherry was more emaciated, Halyburton, also famished, took the larger serving for himself. But when they finished eating, he felt terrible. He didn't say anything but vowed that from then on he would let Cherry select his own meal. Day after day, meal after meal, Halyburton brought in the two bowls and placed them before Cherry. Sometimes Cherry took the larger portion; other times, the smaller. Halyburton concluded that only he was focused on who got more food, that such judgments were irrelevant to his cellmate. He admired Cherry's apparent indifference to such petty concerns.

Cherry, in fact, found Halyburton's handling of the food admirable. He was mindful of the portions but did not want to take more than he deserved. What was more, he appreciated that Halyburton gave him first choice, and he also noticed that Halyburton would not eat until he had finished, just in case he needed more food. On several occasions, the Vietnamese, trying to energize Cherry, left piles of sugar, which could be spread on bread and were considered a delicacy. Halyburton could have taken them for himself, but he never did.

While Halyburton was directly contributing to Cherry's physical well-being, Cherry was having his own effect on Halyburton. Until they met, Halyburton had been interested only in how he was going to survive. As a junior officer, he was low in the chain of command, so he knew he would not be a central figure in the prisoners' resistance. He felt sorry for himself, but Cherry's example began to shake him from his self-pity. Cherry was in far worse shape than he — his shoulder was in severe pain — yet he never complained. Halyburton could not lament his own plight when his roommate seemed to bear his own suffering with so much pride and determination.

Cherry's example prodded Halyburton to try to do more for the Americans' collective resistance. Using the tap code, he asked Knutson the names of all the prisoners he knew so that he could memorize them. He wanted to reach other POWs as well, ask them who they knew had been captured, and keep a running tally in his head. (Other Americans were doing the same thing.) This gave Halyburton a chance to use his mind and stem the boredom, but it was also critical to ensure that, once freed, no one was left behind. Thus, Halyburton committed himself to memorizing every name, an assignment that became increasingly difficult as the years passed and the numbers swelled. (In December 1965 fewer than sixty-five Americans had been captured in the North.)

Halyburton had already met a handful of prisoners at Heartbreak, and now Knutson tapped him other names, sorting them by date of shootdown. "Everett Alvarez . . . Robert Shumaker . . . Carlyle Harris . . ." Halyburton later learned the names alphabetically, then by rank and service, and even by cell. He was no longer in his F-4, but he had rejoined the war effort.

One thing that Cherry and Halyburton shared — indeed, a central part of their time together — was their love of cigarettes. Before they were captured, Halyburton smoked a pack of Winstons a day while Cherry plowed through three packs of Camels. In prison, they were limited to three cigarettes a day, distributed one at a time. It was, according to Halyburton, "the one big event of the day — or actually, three events."

They occasionally received inferior imports or damaged domestic smokes: a stick or tobacco stem would end up inside the wrap, making it difficult to draw, or a loose seam allowed the weed to escape. But compared to the food, the cigarettes were pretty good — strong, aromatic, with a definite kick. Their quality may have reflected the personal taste of Vietnam's most famous smoker, Ho Chi Minh.

Cherry and Halyburton tried to make their cigarettes last as long as possible; but without tobacco additives, the cigarette would go out unless puffed, a serious problem if they didn't have matches. (Stolen matches were valuable contraband.) In later years, many of the prisoners solved this dilemma by making smoldering "punks" out of toilet paper. The cigarettes themselves were occasionally withheld as punishment, and one guard who delivered them in pans would sometimes shake them to knock out the tobacco, which could then be smoked by the Vietnamese.

But cigarettes were elemental to both prisoners, a balm to their senses, a brief remove from their burdens. The soft white wrapper felt like cotton against their fingers, the glowing tip waved in the air like a spark of life. Each man inhaled slowly, holding his breath for a full count of unfiltered pleasure, releasing tides of nicotine into his system and — in Cherry's words — "satisfying every nerve." They blew the smoke out unhurriedly, smoothly, through their mouth or nose, the way serious smokers do.

Smoking was, Cherry said, "the most important thing of the day." He and Halyburton discussed the cigarettes before they arrived — when they would come, what their condition would be. They talked about their quality once they were received and then rehashed the experience when it was over. It was bliss, it was communal, and it was three times a day.

Virtually every POW spent considerable time thinking about his wife or girlfriend, remembering past intimacies, imagining idyllic reunions, and summoning a world with unconditional love, emotional support, and extravagant affection. By sustaining these relationships mentally, the prisoners sustained themselves, giving them a powerful incentive to survive. Their faith, however, was tested by constant fears that their loved ones had moved on to other relationships, fears that intensified with each year in captivity. What was she doing? Who was she seeing? Was she faithful? When a POW learned, either in Vietnam or after his repatriation, that his wife or girlfriend had ended the relationship, it was heartbreaking. One prisoner whose wife divorced him while he was in captivity later said he resented his former wife far more than the North Vietnamese.

For Cherry and Halyburton, the fate of their wives, as well as their children, weighed heavily on them. The men discussed their families often, but while they were both husbands and fathers, their circumstances could not have been more different.

Fred and Shirley had met in 1954 when Fred was stationed in Great Falls, Montana, and Shirley was a hostess at a club. According to Fred, they married after Shirley became pregnant. (She already had a son, Donald, from a previous relationship.) Their first son, Fred Jr., was born in 1955, and two daughters, Debbie and Cynthia, followed. To outsiders, Shirley was considered a dutiful Air Force wife who followed her husband to different postings — Delaware, Oklahoma, Arizona, Japan — raised the children, and participated in all the right social activities, such as a mahjong club on the Yokota Air Base.

But the marriage itself had little chance for success. Shirley later told Cynthia that marrying Fred "was my ticket out of Montana."* Once she was pregnant, Fred knew his career would be hurt if he had a child out of wedlock; his standing, however, might benefit if he were seen as a family man. Marriage seemed to serve everyone's purpose, but only briefly. Shirley resented Fred's infidelity, and she also had a temper. When he came home late, she would wax the floor to make him slip. Fred belatedly concluded that combat pilots were better off single.

Despite the turmoil, Fred remembered the gentler moments with Shirley and worried how she was faring with four children. His daughter Cynthia later speculated that Fred really fell in love with his wife during his years in prison.

Porter and Marty, married for less than two years, gushed affection for each other like young newlyweds. When Porter was on the *Independence,* they communicated through letters and tapes, conferring about a second honeymoon "even if we don't go anywhere," Porter wrote. They dreamed about having a second child — a son, they hoped. Porter told her that in Singapore he bought her a bolt of ivory raw silk and a set of china. "I love you more than words will ever say," he wrote. "In my mind and heart, you

*In an interview, Shirley declined to discuss most aspects of the relationship, though she did confirm that she wanted out of Montana.

are always with me. I miss you, I miss you, I love you . . . My life is pretty grim no matter how busy I keep myself."

Marty tended to her newborn and struggled to make ends meet but was smitten with motherhood. "Did you know that the astronauts use the same soap Dabney does?" she wrote. After Porter sent a photograph of himself on the flight deck, she wrote back, "Where's your ring?" He explained in his next letter that an aviator did not wear it in combat because if shot down and captured, his marital status could be used against him.

In another letter, Marty wrote:

> Julius, I'm so very happy to have you and Dabs. Everyone here has no particular happiness or even knows that it exists. They think I'm very young, naïve, and innocent to be happy and living for something. It may be that happiest characteristic of youth but it isn't something you have to lose. I don't think making you and Dabs happy could ever grow old to me.

The conversations between Halyburton and Cherry reflected their different domestic lives. As a new father, Halyburton asked Cherry about raising children, and somehow the conversation veered into the question of who was more important — your wife or your kids.

"That's easy," Cherry said. "I'd take my kids."

Halyburton couldn't imagine anyone being as important as his wife. "I'd choose Marty," he said.

"After you get to know your kids," Cherry responded, "you grow to love them a lot more."

The exchange reminded Halyburton that he had barely met his daughter and knew her not at all. But in his mind her importance began to grow. Recalling his own childhood, he did not want Dabney to grow up without a father.

*　　*　　*

Cherry didn't know that the Zoo had showers until he met Halyburton, but even then he didn't demand one. It was never his style
to push for things, and experience had taught him to be wary of a
backlash. Halyburton, not so reticent, continually pestered the
turnkeys and guards to allow Cherry to shower. He needed another one as well. Either because of his insistence or because the
authorities just wanted them clean, in the middle of December,
Cherry and Halyburton were sent to a nearby building the Americans called the Pigsty. In addition to about a dozen cells, it had a
shower.

Cherry took off his shirt and sling, allowing Halyburton a clear
look at his injury. It horrified him. The shoulder was simply gone,
replaced by a large indentation. A dislocated bone protruded gro-
tesquely from the sunken mass, and it did not seem connected to
anything else. While the broken bones in Cherry's wrist and ankle
were healing, his moon crater of a shoulder was in constant pain
and on its way to ruin.

"I'm sorry about that, Fred," Halyburton said.

"So am I, Haly, so am I."

The water was cold. Using lye soap and his one good arm,
Cherry washed his hair and most of his body, but he couldn't
reach around to scrub his back. Halyburton saw him struggling.

"Hey, Fred, do you want me to wash your back?"

He wasn't sure if Cherry would let him, but his friend was
grateful. Cherry knew he would have been too complacent to demand a shower, yet Halyburton's insistence demonstrated how
much he cared for him. Cherry silently turned his back, and
Halyburton scrubbed the washrag with soap and began to rub.

8

No Ordinary Prisoner

December 23, 1965, was Halyburton's second wedding anniversary. Though marooned in North Vietnam, he thought the day might bring him luck.

That evening, a guard told Halyburton and Cherry only that they were leaving their cell; no other information was given. Halyburton rolled up both bedrolls, and they were taken to a nearby building known as the Pool Hall. It was their lucky day indeed. Their new room had fresh green and white tile that was easy to clean. Their bunks were elevated on a low brick pedestal, which made them easier to sit on, particularly for Cherry. The walls were whitewashed, making the room brighter. After settling in, they heard the turnkey walking down the corridor, dropping things along the way. Halyburton went to the door and looked beneath. *Thwack!* Falling to the ground were two pairs of rubber sandals, which he retrieved when the turnkey opened the door.

"Fred, I think they brought us my anniversary present," Halyburton said. Green socks arrived a few weeks later.

The sandals not only protected their feet but had an unintended benefit for many of the POWs: they were placed on the ragged edges of their wastebuckets to protect their buttocks.

*   *   *

On Christmas Eve the American bombing stopped, and President Johnson launched a "peace offensive," including a fourteen-point program that invited the North Vietnamese to enter into "negotiations without preconditions." Whether Johnson was more interested in public relations than peace is a matter of debate — Hanoi radio called the offer a "trick" — but Cherry and Halyburton were told during interrogations that bombs were no longer falling. Ironically, such pauses discouraged them, for they believed — like most of the POWs — that they would be released only after the United States scored a decisive military victory. But in this case the lull coincided with Christmas, which prompted the Vietnamese to lighten their treatment. For several weeks the food improved, rations were doubled, and the cell windows were opened to admit some air. Physical exams were ordered for the Americans, and the interrogations were less intense. The enemy was motivated by self-interest as much as humanitarianism. If the war did end, they wanted to present healthy POWs as an example of their magnanimity.

Cherry and Halyburton received a taste of this new spirit on Christmas Day, when they were questioned jointly for the first time. Inside the interrogation room was a small Norfolk pine draped with a string of painted light bulbs, and they were given hard candy, a cup of orange liqueur, and three cigarettes. They were told about the bombing halt and peace plan, which the Vietnamese scorned. But later, for dinner, they received turkey, fresh vegetables, salad, cookies, and fruit. It was the best Halyburton had felt in a long time, and he thought that offering a gift would be the right way to end the holiday.

"Fred," he said, reaching into his pocket for the most valuable thing he possessed, "I don't have much to give you, but here's my last cigarette."

"Well," Cherry said, "I don't have much to give you either, but here's my last cigarette."

Their expectations of being home by Christmas had been thwarted, but they felt confident that they would be released by summer, when they knew the cells would be unbearably hot. Their faith had nothing to do with any evidence about the progress of the war or possible peace negotiations. It was a matter of emotional self-preservation. To believe that their incarceration would last longer than six months would be too depressing to bear.

Before going to bed, the roommates slowly smoked their last cigarettes.

In the middle of January, Cherry was visited by a contingent of officials — a doctor, a medic, a turnkey, and an interrogator. They asked Halyburton to take off Cherry's shirt, and the doctor began feeling the injured shoulder and drawing lines with his finger. It was clear that he was discussing surgery; a robust black prisoner would be evidence that the Vietnamese treated minorities better than white Americans did.

But the prospect alarmed Halyburton. "Fred," he said after they left, "don't let them cut on your shoulder if you can help it. If you're not in great pain, I wouldn't trust these guys. If they do it, the risk of infection is really big."

Cherry was in pain but agreed he'd rather not be cut open in a country with primitive medical care. "I'd rather go with the way it is," he said.

The doctor came back two more times to examine the injury. One time, he used a pen to draw a line across the shoulder.

"Fred, it looks like they're going to operate," Halyburton said.

The Vietnamese continued to pay special attention to Cherry. They placed orange juice and cookies outside his cell to make the other prisoners believe a black man was receiving preferred treatment. The other POWs never bought it, and Cherry never actually got the juice and cookies. While the air campaign resumed on January 31, 1966, ending the prospect that the POWs would be freed en masse, Hanoi still released individual prisoners to score propaganda points, and Cherry's race made him an ideal candidate. At the time, he was vaguely aware of his special status, but on the night of February 9, he knew for sure that he was no ordinary prisoner.

The medic entered the cell and tapped his right hand on his left wrist, indicating that Cherry should put on his long-sleeve shirt. The Vietnamese called this to "dress seriously" or respectfully, the Americans showing respect to the authorities. It was typically demanded before seeing an interrogator, but not tonight.

"I think I'm going to the hospital," Cherry said.

"I think you're right, Fred. I'll be praying for you."

For Halyburton, the first night without his cellmate was lonely, and he prayed for his friend.

As Cherry was taken to a Jeep, blindfolded, he assumed that he would return to the same cell because his belongings were still there. But he wasn't sure, and he wondered who would take care of him if he ended up in some other room. He was driven through Hanoi and taken inside a building; even before the blindfold was released, he could smell the ether and the bedpans and the disinfectant. He was inside a hospital, surrounded by men and women in white smocks. He was then taken to an operating room, placed on a bed, and stuck with a needle. He feared he was being injected with truth serum to force out information on America's nuclear weapons systems. He knew a good deal, so he waited, focusing on an overhead light, and prayed that he would not reveal anything that would hurt his country. Fortunately, the injection was an anesthetic; America's nuclear secrets were safe.

When Cherry woke up several hours later, he found himself in a cast from waist to neck, including his left arm. The plaster was still wet. He was put on a stretcher, hauled back to the Zoo, and returned to his cell, groggy. Halyburton was relieved to see him but aghast at his condition. The wet cast indicated that the hospital had discharged him prematurely, and he looked stiff and uncomfortable. Halyburton helped him to his bunk, rolled up some clothes to make a pillow, and covered him with his blanket. Cherry couldn't move or talk clearly, but he thought to himself that he was grateful to be back — it felt as though he had returned home.

The next morning, Cherry tried to explain what had happened. "They put me on this table, gave me some ether or anesthesia, and I woke up in this damn cast," he said. He was in extreme pain and used his one good arm to pull the cast back. "I can hardly breathe," he said. "It's too tight."

Halyburton slid his fingers beneath the plaster. Instead of padding, he found only a thin layer of cotton to protect Cherry's skin. The cast was not just immobilizing Cherry, it was suffocating him. The surgery had left him faint and all but incapacitated. Whenever Halyburton asked how he was doing, Cherry repeated, "It's hard to breathe."

Halyburton didn't know what to do. He tapped his information to Air Force Captain Quincy Collins in the adjacent cell; Collins could only tap back encouragement. Halyburton prodded Cherry to eat, but the food, by expanding Cherry's stomach, increased the tightness of the cast. He typically pushed away the plate, depriving his body and weakening him further. Perhaps his biggest problem was using the bucket. The cast prevented him from sitting up, bending, or pulling down his shorts. But he couldn't bring himself to ask Halyburton for help, saying instead, "I really have to figure out how to use this bucket."

"I'll help you," Halyburton said.

He lifted Cherry out of bed — in Fred's words, "like a newborn" — slid his shorts down, and held him steady so he could urinate.

That was relatively easy; sitting on the bucket was not. When the time came, Halyburton pulled the bucket over to the side of the bed, then positioned him so one cheek was on the side, the other hanging over. And there, he held him. It was imperfect, but it worked. Cherry could only say "Thank you."

The routine was repeated for many weeks, but with new complications. Cherry began to battle fevers and passed out while he was trying to stand. Halyburton laid him down and put a wet towel across his forehead until he revived. When he realized what happened, his words of thanks seemed shallow. Instead, he quietly wept, hoping Halyburton didn't notice. (He didn't.)

"I was crying about the kindness," Cherry said later. "I tried to restrain it, but I don't mind shedding a tear when there is good reason."

Halyburton feared that inactivity would cause Cherry to wither away, so he wanted his roommate to exercise. "You have to walk to get your strength," he said.

"Oh, Haly, I can't."

And he couldn't, at least not by himself. So he draped his right arm around Halyburton, leaned against him, and the two inched their way around the cell. It was cold, and Cherry began breathing heavily. They had walked for only a few minutes when Cherry, exhausted, grabbed Halyburton, who caught him and carried him back to his bunk, like a soldier leaving a battlefield who would not leave his buddy behind.

"I can't go anymore," Cherry said.

"That's okay," Halyburton said. "We'll do more tomorrow."

And they did, Halyburton waking him up, lifting him, imploring him to walk, and all but carrying him around the room until Cherry's body, at eighty-five pounds, frail and wasted, could no longer stand.

Halyburton did anything he could to keep Cherry awake and

talking, the latter to ensure that he was breathing fresh air into his body. Halyburton recounted novels, movies, and his own life history and asked Cherry about his own family. Cherry knew he was giving Halyburton familiar answers, but he understood his purpose — to keep him alert, conscious. The mind games were appreciated, but of all the things Halyburton did, his insistence on dragging Fred's limp body around the room made the most indelible impression. "He was trying to keep me alive," Cherry later explained.

Halyburton's respect for Cherry was such that he did not see his help as a sacrifice. "It was a privilege," he said.

If Cherry believed that Halyburton was different, he was not alone. Porter was different, though those who knew him in his youth often use another word to describe him and his family: exotic.

His grandfather had a lot to do with the family's unconventional image. Born on an Iowa farm in 1874, he was sent by the Presbyterian Church to India, where he taught college science for six years. At home in Davidson, he displayed rare Indian artifacts, like a small ivory elephant nestled inside a bean and a sword stained with human blood. He imported ginkgo trees from China, erected a bamboo grove in the backyard, raised rabbits in a hutch, and maintained feeders for birds.

Porter's mother, Katharine, was even more atypical, a single mother with a career in a town where most mothers were married and did not work outside the house. She wrote for the Society pages of the *Charlotte Observer* and was named women's editor in 1952. Her important position fit her striking image. She had brown-reddish hair, a coy smile, and, in the words of one admirer, a "slinky walk." She gestured elegantly with her hands and enjoyed sunbathing, her olive skin tanning well. In a community of staid fashions, she wore bangle bracelets, costume jewelry, and bright scarves, belts, and hats; nail polish, powders, and perfume were all applied generously.

She was fascinated by the occult and considered herself a mystic, claiming to have had an accurate "vision" of a former boyfriend dying in battle in World War II. She enjoyed ballroom dancing and bridge, and when she played the piano, the rings on her fingers would glint in the light as they raced across the keyboard. When she returned to Davidson to spend more time with her aging parents, she worked in the college's alumni and public affairs office, filing photographs, press releases, and other materials in such a way that no one else could find them.

When she and her husband were divorced a few years after Porter was born, Katharine made one demand: he could not see or talk to Porter again.* Katharine protected her son fiercely and taught him skills that most boys did not have — how to play bridge, dance, cook, and sew. When Porter dated a girl from the "wrong" side of the railroad tracks, where working-class whites and blacks lived, Katharine demanded that Porter end the relationship. In high school, when Porter rebelled and became a self-described hell-raiser, Katharine sent him to Sewanee Military Academy in Tennessee, where he had to make his bed tight as a drum, fold his shirts ten inches wide, and march each Sunday in a gray wool uniform and shiny black shoes. The school was expensive, but Katharine would have paid any price to ensure that her son received proper discipline and training. Porter was grateful for her sacrifice.

Porter was like his mother in some ways. He shared her passion for writing, the arts, and bold fashions. Inspired as a teenager by Elvis Presley, he wore a slick pompadour, a long-sleeve gold and

*Porter would not meet his father until after he returned from Vietnam.

black shirt, and tan slacks with narrow cuffs, star-shaped pockets, and belt loops with pearl snaps. Lanky and handsome, he played sports, fished, hunted, carved wood, and made friends easily. His world had a veneer of structure and order, his comic books neatly stacked in his bedroom, his clothes washed and ironed by a maid, and his grandfather dressed in a tie at dinner.

But Porter deeply felt the absence of his father. When other fathers took their sons to cub scouts or boy scouts, he went alone. Though he rarely discussed his family, his friends could tell that something was amiss. Porter could also be aloof, moody, or melancholy, at times given to fits of rage or depression. "He was a nice guy and he was more in touch with his emotions," recalled one friend, Marie Fisher Bjorneboe. "But there was a sadness to Porter."

He was no choirboy. He smoked and drank at thirteen and even learned how to make home brew. His rebel image was enhanced by his dexterity with a bull whip: he could snap a cigarette out of someone's hand at fifteen feet. He harassed neighbors with crank phone calls, shot out streetlights with a BB gun, and lit fireworks on the Fourth of July — the police arrested him for disturbing the peace. His Elvis impersonation, as well as his continual flirtations with girls, was considered scandalous.

He found more constructive outlets through reading and writing. Among his favorite books were those by Richard Halliburton, who in the 1920s and '30s wrote about climbing the Pyramids in Egypt and Mexico, floating on the Great Salt Lake, and excavating ancient Pompeii. The tales excited Porter, all the more so because his mother told him that Halliburton was a relative, though his name was spelled differently. Porter took the books to school and told his friends that Halliburton was his uncle.

Porter was a determined writer. He spent one summer in high school sitting on his porch, trying to write a novel on his Royal typewriter. It was called "Magyar Mansion," not because he liked Hungary — he didn't even know what "Magyar" meant; he just liked the exotic sound of the word. He wrote his first poem in eleventh grade, after the death of Dylan Thomas. Another poem, to a girlfriend, read in part:

> Who comes back to haunt my dreams,
> Who comes back to stir
> A thousand little memories flying back to her,
> Not but one could do so much,
> And wet my eyes with tears.
> Looking back to happy times,
> Back across the years.

In 1959 Halyburton entered Davidson College, the prestigious male sanctum that exalted the Christian ideals of faith, honor, and service. The students were required to take courses in both the Old and New Testaments, to attend chapel service three times a week, and to be present at vespers on Sunday evening.

The college had about 850 students, and in the early 1960s they would stroll across campus carrying their books in suitcases and wearing khakis, Madras shirts, and loafers without socks. Most came from the South and were ambitious to excel in business, law, government, or medicine — young men with a sense of their own destiny. The campus itself emphasized order and unity through its neoclassical architecture and design.

Its traditions assumed that the students were men of honor. The library, for example, did not mandate that books be checked out but trusted they would be returned. Expectations were specified by the school's sacred Honor Code, which every student learned during a lecture on his first day of school. There was a loose, unwritten code to haze freshmen. A first-year student had to say hello to anyone he saw on campus, was forbidden to walk on the grass, and had to wear a beanie. Failure to comply might require him to scrub the dormitory floors. But violations of the written code were punished severely, with several students usually expelled each year. The expulsions themselves were public events. The announcement was made at chapel, though no name was mentioned, causing the students to swing their heads around and search for faces to determine who was missing. The fallen student was soon identified, and the message was clear: if you trespass, your disgrace will be recognized by all your peers.

The most serious offenses were plagiarism, cheating, and stealing, so on every term paper or assignment, a student had to pledge in writing that the work was his own. If he broke that pledge, his fate was decided by a student Honor Court.

For all the import it gave to matters of honor, Davidson was hardly devoid of fun. Halyburton bypassed an appointment to the U.S. Naval Academy for the more open life at Davidson. For three years he was the social chairman of the Phi Gamma Delta fraternity, organizing beach parties, reaching out to women's colleges, and generally ensuring that the boys had plenty to eat and drink at any social event.

But his college years were about more than carousing. The school, with its liberal arts curriculum, nurtured his philosophical bent. While he was not partial to all the religious services, the discussions of beliefs, spirituality, and higher truths appealed to his own search for understanding, and Davidson's ethos of discipline and piety — the idea that life was a series of moral choices that determined one's destiny — gave him lasting guidance.

Caring for Cherry wasn't Halyburton's only job. Prisoners were often assigned work outside their cells, such as sweeping sidewalks or roads, chopping wood, even making coal balls. During much of his time with Cherry, Halyburton washed dishes, which he enjoyed. It got him outside and allowed him to do something useful.

The bedridden Cherry relied on Halyburton to be his eyes to

the compound, and Halyburton, who could see through the slats in the window, sometimes turned his reports into a game. There was the interrogator who supervised the camp's reading materials and always carried a book. While the other prisoners called him Dum Dum or Colt 45 (for the revolver he also carried), Cherry and Halyburton called him Shakespeare.

Halyburton saw John Frederick, a brawny Marine warrant officer whose hands had been burned badly during his shootdown and who had been blindfolded and in leg irons for a month. He described Frederick's limping around the compound by himself, his fingers appearing "as if they've grown together." Word had spread about his mistreatment, and Cherry thought about the pain he had endured, perhaps like his own — except he had a roommate. Frederick later died in captivity from typhoid.

As the days passed, Halyburton tried to relieve the discomfort of Cherry's cast by gripping it along the edges and lifting it — in effect, allowing Cherry to breathe — for an hour a day. He also brought back from the washroom bowls of water and, using the small towels used for washing dishes, gave Cherry a sponge bath, cleaning his face, arm, and legs.

These efforts gave Cherry some temporary relief, but Halyburton knew that he was growing worse. One day, while he was lifting Cherry over the wastebucket, he tried to pull down his shorts, but they were stuck against the backside. He tried again, but they still wouldn't move. Then he saw a dark spot at the site of his tailbone and realized that a bedsore had bled through and fused the shorts against his body.

"Oh, God, Fred, you've got a big hole over your tailbone from lying on that bed."

"Yeah, I could feel that hurting," Cherry said. "It's uncomfortable." To Halyburton's amazement, he had never said anything about it.

Halyburton placed Cherry's blankets beneath him so his tail-bone wouldn't rub against the bed. He also demanded to see the medic — "*Boxi! Boxi!*" The one who appeared didn't speak English, so Halyburton used hand gestures to explain Cherry's injury. The medic, scowling, said no, and left.

Cherry discovered his next problem himself. Just as the bunk wore away his skin, his cast was having the same effect on his wrist. Halyburton again demanded to see the medic, who saw the problem and this time returned with surgical scissors. He gave them to Halyburton, who cut back the flimsy plaster. But the abrasion was only a hint of what was really happening. One day when Halyburton was lifting the cast, he heard a slurping sound. Fluid had gathered beneath the plaster. Halyburton assumed, correctly, that pus was flowing from the incisions, which had become infected.

Halyburton knew that the cast was killing Cherry, and he demanded to see not the hapless medic but an interrogator — "*Bow cow! Bow cow!*" — who spoke English. Halyburton told him that the cast was hurting Cherry and his body had become infected.

"He'll die if you don't help him," he said.

The interrogator returned with the medic, who again brought his scissors and told Halyburton to cut the cast at the elbow. When he did, a pint of thick, malodorous, green-yellow pus flowed out. The stench was so bad that the medic, interrogator, and guard turned around and left the cell. The medic came back with a bucket of water and rags and told Halyburton, "You clean up." As he opened the door to leave, Halyburton stepped toward him.

"No, no! Cherry will die if you don't help him and do something about this cast!"

The medic looked at him, walked out, and shut the door.

113

For Halyburton, Cherry's death seemed like a real possibility. Before, Cherry could shuffle around the room by leaning on Halyburton, but he couldn't even do that now. He simply lay on his back, a complete invalid. To ease the pressure on his tailbone, Halyburton used his own clothes as a doughnut for Cherry to sleep on. Pus soaked the material, but Halyburton could scrub it out with soap and water each day when he washed the dishes. The room stunk, and there was nothing to do about it.

Drifting in and out of consciousness, Cherry felt himself leaving his body and doing what he loved to do most — flying. He didn't fly just any plane. He piloted a B-58, a light, high-altitude jet that held many world speed records but was limited in range and bombing capability. It had not been used in Vietnam and Cherry had never flown it — until his hallucination. "We have B-58s in the war now," he whispered to Halyburton. He described how he had sneaked out of the prison on a secret mission and bombed roads in North Vietnam. "They gave me an air medal, but I told them to give it to Jerry Hooper, another guy in the squadron," Cherry said. "I told 'em I have enough."

Halyburton played along. "Is that right?" he said.

"Yes. The war will be over soon."

Cherry wasn't eating but burned his own fat for energy. He felt as if his body were eating itself. He developed a fever, and as his temperature rose, his delirium became more prolonged and his dream sequences more bizarre. In one, he sat in a greasy-spoon restaurant in South Vietnam, hungry, and a woman was frying pork chops. The steam was rising from the grill and it was getting warmer, and the steam kept rising. Then a small man approached Cherry and said that he could take care of air conditioning. Cherry said he was a prisoner in the North and that he'd been having problems with his air conditioning. The man said he would see if he could help.

114

Cherry then opened his eyes. He was awake but still addled and feverish. He tried to focus his eyes, and he saw two little men, about a foot high, standing on his chest. They were dark, with oversized eyes and heads, and they were working on the lower half of his body . . . turning screws . . . pressing buttons — they were working on his air conditioning! "Please, hurry," he told them. He just wanted to cool down before his fever killed him. Sometimes they seemed to make progress and his temperature would drop, but the work was slow. He looked up at Halyburton.

"Have you seen them?" Cherry asked.

"Seen who?" Halyburton said.

"The little men."

"What little men?"

"The little air-conditioning men. They keep me cool inside."

Halyburton tried to lift his cast up.

"Maybe they went for more fuel," Cherry said.

"Fuel?"

"They keep me cool. I hope the guards don't catch them."

"How do these men get in here?"

"They come in under the door. They work on my air-conditioning unit."

"You're right then," Halyburton said. "They probably went for more fuel."

His words of comfort would stay with Cherry for years to come. Halyburton could have said there were no little men — or told him that he was crazy — but he didn't. When Cherry was on the brink, his roommate gave him hope that his air conditioning might be fixed.

It was clear that surviving Vietnamese medical care might be as challenging as surviving the torture. Once, a doctor came to the cell to give Cherry a shot, bringing along a burner so he could sterilize the needle in boiling water. He pulled it out with forceps, dropped it on the ground — then attached it to the syringe.

"You need to put the needle back in the water!" Halyburton yelled.

"No, no, no," said the doctor, waving him off.

Halyburton kept demanding to see the medic, warning the guards that Cherry was going to die. He got no response. Then, on March 12, more than a month after the cast had been put on, the guards moved Halyburton and Cherry to another section of the Zoo, a building called the Garage. There was no explanation for the move. The new cell was larger but darker, mustier, and less comfortable. Their bed boards were on sawhorses.

Cherry, semiconscious, knew he was near death, but when he was lucid he always had the same thought: I will not die in this prison, I will not die in this prison. He had fought too hard all his life to die in this prison. He also realized how much he depended on Halyburton, how he couldn't survive without him. He never wanted to leave him.

On March 18, the medic came to the cell and had Cherry taken out on a stretcher; his belongings remained. He and Halyburton believed he'd be coming back, but when the other prisoners saw him leave, word spread through the camp: Cherry had been taken away to die. Actually, he was returned to the hospital, once again for a short visit. As the cast was removed with a scissors and a saw, Cherry saw his decaying flesh fall off. His thighs looked the same width as his arms. He figured he weighed about eighty pounds, skin wrapped in bone. He also had lagoons of pus on his body and at least a half-dozen bedsores. Then someone picked up a green beer bottle and poured it over his torso. The bottle contained not beer but gasoline, which burned like fire over his gaping wounds. The fumes overwhelmed him.

The gas was not for torture — it wasn't the "season" for torture, according to Cherry, and the authorities had more devious instruments at hand. It was considered a healing agent, but its vapor was so strong that it knocked him out. He awoke to the doctor slapping his wrists, leading him to believe that his pulse had stopped. He was then given a blood transfusion and fed intravenously.

Returning to his cell in bandages, Cherry was carried in on a stretcher, which was placed on his bed board. An intravenous tube went into his arm, its bag of fluid held by a stand. His bad shoulder was back in a sling, his fractured wrist still not healed. Halyburton draped the mosquito net over him, exposing only his arm with the IV. Cherry had been gone only a few days, but Halyburton was stunned by his physical deterioration. Asleep, he looked more dead than alive.

The next morning Cherry complained that his shoulders were being crushed. Halyburton discovered that the stretcher was supposed to be elevated by six-inch legs, but the bar to lock the stretcher and keep the patient elevated wasn't working, so its sides were pressing in against the sunken Cherry. He felt as if he was being squeezed to death by his own body. Halyburton piled extra clothes beneath Cherry's shoulders, easing the pressure.

"We're in a helluva mess now, Fred," Halyburton said, a line he'd occasionally use to lighten the mood.

Cherry described the removal of the cast and the gasoline bath. Halyburton just shook his head as he said, "I can't believe they did that."

Though now free of the cast, Cherry was in no better shape. He still couldn't move his arm, and all he had to show from the cast was a body pocked with sores — Halyburton counted nine.

At least he was now receiving antibiotics and was no longer hallucinating, but the IV was no panacea. Unknown to Cherry, in the hospital a small tube had been inserted in his ankle, which could be hooked up to an IV line. Presumably this approach would spare a medic the difficulty of finding a vein in his desiccated arm, making transfusions easier. So the medic came to the cell, disengaged the IV, and inserted the line into the ankle tube. After he left, Cherry looked up at Halyburton and said, "Bubbles are going up my leg."

Halyburton thought it could kill him, so he banged on the door and yelled for an interrogator. Not willing to wait, he pulled the line out of the tube, causing the fluid to pour onto the floor.

The medic came first, and Halyburton, frustrated and angry, pushed him toward the door and demanded to see a doctor or interrogator. A guard retaliated and slammed Halyburton against the wall, but he got what he wanted. Eagle, the camp commander, came to the cell. Halyburton hadn't seen him since Heartbreak.

"There's air in there!" Halyburton said, pointing to the tube. "You're going to kill Cherry!"

According to Eagle, Halyburton should have been grateful that the Vietnamese had saved Cherry at the hospital. "Cherry almost died," he said.

"Yeah, that's what I've been trying to tell you for a long time," Halyburton said.

"Yeah, but you have a very bad attitude. You don't speak to guards and doctor that way."

The following day another doctor came to the cell, cut an incision in Cherry's other ankle, and inserted a tube. This one, however, was never used, and soon both tubes were removed and the ankles bandaged, leaving two scars as permanent reminders of Vietnam's medical care.

The new shoulder bandage created a different problem: it was too tight. "My arm is going to sleep," Cherry said. Halyburton investigated the wrap and decided he could do better. He unwound the bandage and for the first time saw the impact of the cast: two

As an aviation cadet in 1952, Fred Cherry was often the only African American among his peers, but his piloting skills erased conventional prejudices against black fliers. Courtesy of Marion Godwin

The cockpit was Cherry's ultimate refuge, allowing him to fly above the poverty and segregation of his youth.
Courtesy of Marion Godwin

Commissioned in 1952 as a second lieutenant, Cherry became a pioneer in the integration of the Air Force.
Courtesy of Marion Godwin

During the Korean War, Cherry performed a daring airborne maneuver when he used his wing tip to secure the landing gear of another jet.
Courtesy of Marion Godwin

With his adopted son, Donald, in 1955; Cherry was
known for his perfectly creased clothes and sense
of style. Courtesy of Marion Godwin

In 1973 Cherry accepted his first light as a free man,
at Clark Air Base in the Philippines.
Courtesy of Marion Godwin

In 1981 the Air Force commissioned this portrait of Cherry, which now hangs in the Pentagon.

Portrait by Harrison Benton

Cherry's sister, Beulah, raised him as a boy and was his principal advocate when he was a POW. Here they stand with Air Force Colonel Clark Price during Cherry's return to Suffolk, Virginia. Photo by the *Virginian-Pilot*

Katharine Halyburton, a fiercely protective single mother, taught Porter skills such as playing bridge, dancing, cooking, and sewing.
Courtesy of Porter Halyburton

Commissioned as a Navy ensign in 1964, Halyburton was grateful that the Vietnam War would give him an opportunity to see combat.
Courtesy of Porter Halyburton

Halyburton (left), standing on the flight deck of the USS *Independence* in August of 1965, admired his pilot, Stanley Olmstead, whose good looks, humble roots, and aeronautic savvy seemed lifted from a military recruitment catalogue. Courtesy of Betty Dyess

For more than seven years, Marty Halyburton raised Dabney by herself. She sent this photograph to Porter when he was in captivity. Courtesy of Porter Halyburton

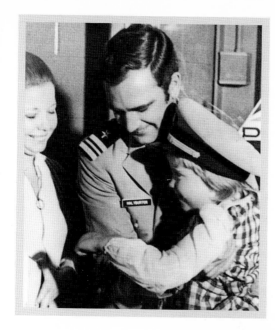

Porter, Marty, and Dabney arrived at Atlanta's airport on March 9, 1973.
Associated Press

Supporters returned their Porter Halyburton POW bracelets, which were converted into a chandelier.
Courtesy of Porter Halyburton

IN MEMORY OF
PORTER ALEXANDER HALYBURTON
LIEUTENANT, JUNIOR GRADE, USNR
JAN. 16,   OCT. 17,
1941   1965
KILLED IN COMBAT OVER NORTH VIETNAM
SON OF KATHARINE PORTER HALYBURTON
HUSBAND OF MARTHA CARRELL DUERSON
FATHER OF DABNEY LORIMER HALYBURTON

Halyburton's gravestone now rests in his backyard. Photo by James S. Hirsch

Halyburton and Cherry were honored at Davidson College in 1986.
Courtesy of Davidson College

long, "angry red" incisions cut across Cherry's shoulder. It appeared that the stitches had been removed, but fluid was leaking through the scars. The shoulder itself looked no different than before the surgery — which is to say, there was no shoulder. Where there had once been muscle was now a deep gorge, and Cherry still had no movement in his arm. It was evident to Halyburton that the doctors had tried to "reconnect" the arm to the shoulder but failed. He could also smell the gasoline.

Cherry's toughness amazed Halyburton; it was a toughness born of many early hardships. He was not going to die in Vietnam because he had had plenty of experience surviving.

The Cherrys' small wooden house, on a dirt road in the middle of corn and cotton fields, had no electricity or plumbing and little space, forcing several children to sleep in the living room. The roof leaked so badly that Fred would have nightmares of porous ceilings for the rest of his life. The family bought hundred-pound chunks of ice, chipped pieces off to refrigerate their food, then wrapped the rest of the block in newspaper to slow the melting. A coal stove warmed the kitchen in the winter, but there was no refuge from the summer's heat. The closest source of fresh water was a spring a half mile away, and Fred, riding his bike, carried home a splashing pail in each hand. When he woke up on Christmas morning, he found his own unwrapped shoebox with that year's treats: raisins, peanuts, an orange, an apple, maybe a little truck. Fred, knowing the family's meager resources, was grateful.

With inadequate water, sewage, and plumbing in and around the town of Suffolk, disease was rampant. Tuberculosis, pneumonia, heart attacks, and infant deaths occurred at alarmingly high rates — two of Fred's siblings died as infants. Only thirty percent of black children in the area met the minimum health standards, according to a 1930 survey by the Virginia Department of Health.

Dental care consisted of yanking out a rotting tooth, using black pepper to numb the pain.

Both of Fred's parents were descended from slave families in North Carolina. His father, John, was one-quarter Cherokee and looked like an American Indian, with swarthy skin, straight coal-black hair, and a flair for storytelling. A laborer and farmer, he was always juggling several small jobs while looking for something better. He would wake up at two A.M. and walk ten miles to a fertilizer factory. Fred knew if his father came home early, he hadn't been hired that day; if he came home late, he'd worked. Either way, he did the same thing the next morning. After a while, the factory gave him a regular job, and he rode to work with another employee, who had a car.

John also owned ten acres of land that grudgingly yielded corn, peas, beans, and potatoes, and Fred got an early taste of farming. At the age of five, he stood on his tiptoes to reach the crossbar of the plow as his father guided the family mule across the dry, hot fields; Fred hoped their work would produce a surplus. On those occasions, the entire family would shuck peas and beans in the evening, pour them into pint-sized baskets, and prepare them for sale. The following day John would lift Fred out of bed at three A.M., wrap him in a burlap sack, and lay him on a mule-driven cart that now carried the food. On the four-mile trek to Suffolk, Fred would wake up beneath the stars and sit next to his father, sometimes eating strawberries that his mother had packed. A lantern attached to the cart illuminated the bumpy roads. Reaching Suffolk by daybreak, John would go from house to house, selling his baskets, as Fred rode the mule. If they sold their entire supply, they earned about five dollars; to celebrate, father and son would stop on the way home for a cookie or a soda and talk about what they would do on the farm that day.

John Cherry died from hepatitis when he was fifty-two; Fred

was only eleven. No pictures of him survived, so Fred, as an adult, did not have a clear memory of his face. The death jolted Fred, who could not imagine how his family would survive. In the end, the tragedy heightened the role of the two people most responsible for raising him — his mother and his oldest sister.

His mother, Leolia, barely five feet tall, was known as "Miss Doll," a name that belied her strength and stamina in raising her own eight children plus a half-dozen grandchildren. She ran down chickens in the coop, then killed, cleaned, and prepared them for cooking. She worked many hours in the field, planting and pulling fruits and vegetables; then cleaning, preserving, and canning them. The family raised hogs, but when the smokehouse held no meat, she mixed chicken's feet into the rice for flavor. She made her own soap by boiling pig fat in a huge iron pot, mixing it with lye, allowing it to solidify, and cutting it into small blocks. At church, she used a washboard as a tambourine.

Even in her sixties, she would race her teenage grandchildren across a field. She could read, though she never attended high school and never learned to drive a car. Her Apostolic faith, whose followers believe in the literal word of the Bible, shaped her life. Almost daily she attended a church service, prayer meeting, or Bible class. Because she believed that a modest appearance reflected pious devotion, she always wore dresses that fell below the knee and shunned makeup and jewelry. Dancing was forbidden; drinking, sinful; personal sacrifice, sacred. She did not buy new underwear for herself so she could save for Fred's college education.

But she was also a tough country woman who punished her children for misbehaving, such as talking disrespectfully to an adult, cursing, or coming home late. She beat them with a switch or whatever was handy, once pulling an entire bush out of the ground and using the roots. "Better not run from me," she'd threaten the guilty party. She also kept a long-nosed, black .38 revolver in a paper bag, locked in a closet, and she was not afraid to fire it. If she thought she heard moonshiners in the woods or chicken thieves around the coop, she'd blast away. She also brandished the gun to enforce discipline. One teenage granddaughter, Joyce, living with her in the 1960s, attended a James Brown concert with a young man without asking for permission. It was a clear violation of the rules; the girls could not be with boys without adult supervision. Leolia grabbed her paper bag, marched to the concert, and found her mortified granddaughter. "Okay, let's go," she said. There was no need to show the gun. Leolia was in her seventies at the time.

If his mother gave Fred strength, endurance, and character, his oldest sister, Beulah, provided him with opportunities and fueled his ambitions. The two siblings always had a special relationship. When Fred was born, his mother handed him to Beulah, who was nineteen, and said, "This baby is yours." Leolia still mothered her youngest child, but she needed help with so large a family, and Beulah was pleased to dote on her little brother.

While Leolia bore the stamp of an earlier time, Beulah was thoroughly modern, a tall, sturdy woman who wore makeup, owned jewelry, and loved to drive. She was the first in the family to attend college (Fred would be the only other one) and was the only child to earn a master's degree. She worked in the Suffolk public school system for thirty-seven years, first as a teacher, then as a principal, and whether she was in school or at home, she demanded the use of proper grammar and respectful language.

She married well. Her husband, Melvin Watts, came from one of the area's wealthiest black families. They operated a bus service, owned part of a popular beach in the Tidewater area, and had other real estate interests. One brother was a doctor and one was a lawyer, while Melvin was a successful farmer. Like Beulah's own father, he was a truck farmer — except Melvin owned at least four trucks, plus a tractor, and would sell his bounty of collard greens,

string beans, radishes, spinach, watermelons, and cantaloupes to
the merchants on High Street in Portsmouth. He bought Beulah a
comfortable home in Suffolk with comforts she'd never had before, such as a telephone, lamps, a refrigerator, and upholstered
furniture.

The couple never bore their own children; they had a foster
daughter, and they cared for their many younger siblings, nieces,
and nephews. Beulah's strict rules and tough love were not welcomed by all. When a nephew fell behind on his mortgage pay-
ments, he asked Melvin if he could borrow $3,000. Melvin agreed
but said Beulah had to give her consent. She refused. "It shouldn't
have happened," she told him. He lost the house.

The bond between Beulah and Fred solidified when he was
thirteen and became bedridden with a mysterious illness. Beulah
volunteered to have him stay with her, where access to a telephone
and car could help in an emergency. On the way to her house,
Fred's body folded in pain and he stopped breathing. Beulah raced
him to the hospital, where he was treated for acute appendicitis.
When he was released, he went to Beulah's home, where he lived
until he left for college.

The move meant one less mouth to feed for Fred's mother.
More important, it put Fred under the supervision of two educated, financially secure guardians. Beulah did not want Fred to
follow the path of his older siblings, who earned a living in various blue-collar or clerical jobs, consigning themselves to margin-
ally better lives than those of their parents. Like his mother, she
made him work hard, waking him up before dawn so he could
load Melvin's watermelon truck, ride into town, and sell the
produce to merchants. Beulah also pushed Fred to improve his
schoolwork, discipline his mind, and prepare for college. She
didn't badger him about his future but simply referred to it.
"When you become a doctor . . . ," she would say.

Fred didn't complain. For the first time he had his own bed, dressed well, and was one of the few students at East Suffolk High School who had regular access to a car. He was also assured of regular paying jobs on Melvin's farm or in other family businesses. While he enjoyed some of the work — he sold beer one summer on the family's beach — he loathed the long hours on dry fields beneath the hot Virginia sun. Picking potatoes, string beans, and strawberries was monotonous labor, frustrating his desire to accomplish something significant. Pulling cotton was even worse because the thorns punctured his hands, causing him to develop the same calluses that both his parents suffered from. He enjoyed driving the tractor, but Melvin sprayed his fields heavily with insecticide, which mixed with the dust that would swirl around Fred's face and penetrate his nostrils. When he blew his nose, the mucus was black.

Fred knew he didn't want to end up like so many of the sun-hardened farmhands around him, with coarse palms, stooped shoulders, and few pennies to save. He had learned how to hunt squirrels and trap raccoons and rabbits but knew it was no livelihood. And even though Melvin had enough money to send him to college, he did not want to be a doctor. Not at all. But he did have another idea, a notion more fantastic than anything even Beulah could imagine, a dream that, if realized, would be his ticket out of the Great Dismal.

In 1936, when he was eight years old, Fred was standing in a cornfield when he heard a rumble in the sky. He looked up and saw a plane descending smoothly in the distance. It was a mesmerizing, inspiring sight — the power and grace of a machine soaring effortlessly above a world where even a sputtering car was considered a mechanical wonder. Fred discovered that the plane landed in a nearby Navy auxiliary airfield, so he began looking for more and tracking their flying patterns. He trained himself to listen for the engines and amazed his friends with his ability to announce that a plane was coming before anyone could hear or see it. In school, he began making paper airplanes and winging them across the room.

After America entered World War II, the training flights accelerated, and Fred stopped whatever he was doing when he heard them. He watched the planes fly in formation, then peel off in mock battles, then twist and turn, the sun glinting off their glass cockpits. The "big birds" flew so low, a mere hundred to two hundred feet above the ground, that the pilots would wave at Fred, who would wave back. He wanted to jump up and get right in; instead, he would race after the aircraft, running through the woods, toward the airfield, then lie in the grass and watch the planes land and take off. "I'm really going to do that someday," he told a cousin, who stared at him in disbelief.

Plane crashes were common. Fred saw some himself, and he knew about others from the flatbed trucks that would cart pieces of the fuselage, wings, and other debris past his house. But the specter of death didn't bother him; if anything, it made the whole experience more exhilarating.

He shared his dream with few people, fearing others might tell him straight out that he was wasting his time, that no colored boy was ever going to be allowed to fly such a machine. He began searching for newspaper and magazine articles about airplanes and combat flying, but only after the start of World War II could he confirm that his dreams of being a black combat pilot were valid. He read about the famed Tuskegee Airmen, the black pilots whose exploits drew extensive coverage in the black newspapers. He closely tracked their every move in the weekly *Afro-American* and the *Norfolk Journal & Guide*, which described the bomber-escort and ground-attack missions in North Africa, Italy, and Germany. Fred kept grainy pictures of planes and studied their curves, dimensions, and capabilities.

He finally shared his ambition with his brother James, who was in the Navy, stationed off the coast of New Guinea. In his letter, he drew a picture of the Curtis P-40L War Hawk, which could fly 350 mph. It was armed with six fifty-caliber Browning machine guns and carried bombs and external fuel tanks. "This is what I'm going to be flying one day," he wrote. In fact, he wanted to be flying it right then. He already knew what kind of planes he wanted to command. Bombers didn't interest him. They flew straight and level; the pilots opened a hatch and the bombs fell out. It was predictable and dull, like driving a bus. Fred craved air-to-air combat, diving and swooping, making split-second decisions, firing bullets, eluding the enemy, soaring straight up and then down, and doing it all alone, without interference from anyone else, master of his own fate. And there was one other thing he wanted — to be the best.

At fourteen, he got his first chance to fly. Near him lived a white man who owned a Piper Cub, a light two-seater, in which he carried passengers for two dollars a ride. Approached by Fred, he expressed surprise at the boy's interest but said he would take him up if he had the money. Fred begged Beulah, who was initially skeptical but by now knew of her brother's ambitions. "I'm just so tired of you worrying about this flying thing," she said, and surrendered the two dollars.

On his first flight, Fred sat in the back seat and strapped on the belt. The man started the engine by twirling the propeller, then hopped in the plane. He maneuvered it across the airfield, reached 30 mph, and pulled the stick. The plane rose and Fred entered a new realm. "I looked down," he later recalled, "and there wasn't a building as tall as I was. I was above everything and everybody, and it made me feel good."

Cherry's wounds were not healing, and on April 10, he was again taken from the cell. His mosquito net, bedroll, and other belongings remained, but this time he did not return the next day or even the next week, alarming Halyburton. He assumed Cherry had been taken to the hospital again but couldn't be sure. He asked the guards what had happened but never received a straight answer. Figuring Cherry's risk for infection was still high, he worried about what kind of treatment he was receiving. His loneliness grew worse.

Every prisoner passed time differently. Some followed the animal life in the cell — the feeding patterns of a spider, the mating habits of a gecko. Others sought more physical activities. One POW crumbled a leftover bandage into a ball and began tossing it up in the air, counting five thousand catches within a few days. Another prisoner estimated that he jogged the equivalent of seven miles a day. Some spent hours just watching a shadow pass across the floor. The combination of physical abuse and isolation led one POW, Air Force Lieutenant John "Spike" Nasmyth, to write, "The POWs who were captured in my era, 1965 through 1968, all went a little crazy after a while."

While waiting for Cherry, Halyburton did a bit of everything. He increased his tap code conversations, he tried making his cigarettes last longer, he walked longer and more vigorously around his cell, sometimes until he passed out from exhaustion. He also tried being as neat as possible, not only in sweeping his cell but by folding his clothes inside his blanket. The Vietnamese had told the Americans how to fold their belongings, but Halyburton wanted to show them that he could do it better, that they were not slobs, and that military discipline and precision still existed in this sinkhole. It was only a blanket and a few garments, but he spent hours striving for the perfect crease.

He was taken to the prison "library," a cell with mostly Communist and pacifist literature, including the writings of Marxists like Wilfred Burchett and Felix Greene. Occasionally, American magazines appeared with antiwar stories. Though desperate for reading material, Halyburton was repulsed by the offerings and refused to return.

For no apparent reason, he was taken to another cell in the Zoo, and he feared he wouldn't see Cherry again, whose belongings remained in their old cell. He was devastated, and he stewed for several days. Then, again without explanation, he was returned to his old cell, where he was reunited with Cherry's clothes if not with the man himself. He continued walking, tapping, thinking, smoking, and folding, but his concern grew over his missing mate.

As Halyburton suspected, Cherry had indeed been taken back to the hospital, for the Vietnamese were determined to improve his condition. His shoulder was infected badly, oozing pus, so he again had surgery. While he was lying on the table — and before receiving any painkillers — his swollen left wrist hurt terribly, so he motioned for help. The doctor saw the problem and drew an X on the wrist. He then picked up a scalpel and cut it, splattering blood.

After the operation, Cherry received antibiotics, had blood tests, and was given fresh bandages. His wounds began to heal; the soft mattress and clean sheets caressed his many sores, and the hot water seeped into every pore on his face. But he couldn't move either arm — they were both bandaged — so he was still helpless. A guard stood outside his open door and was responsible for shaving him (which was unpleasant) and bringing in his food (which he would delay). The tedium, however, was broken each night when two teenage girls cleaned the room. They moved gracefully, sweeping the floor and emptying baskets, and Cherry smiled at them. They spoke no English, so they waved back and smiled coyly. One night they brought him hard candy, unwrapped it, and tentatively walked over to him. If anyone saw them, they'd be in serious trouble. Unable to accept the treat with his hands, Cherry opened his mouth, and one of them dropped it in. It was the sweetest candy he'd ever had. The following night the girl brought a banana, unpeeled it, and tried the same routine. But this time the girl, fearing she'd be caught, shoved the fruit in so quickly it almost choked him. It was less satisfying than the candy, but he appreciated the effort.

Mostly he lay quietly in bed. But he kept his mind active.

Most POWs were forced into long periods of fierce introspection; one writer, Geoffrey Norman, distinguished between "idle daydreaming and disciplined fantasy." While the former consisted of recalling random events, the latter had more structure and purpose. Jack Fellowes, for example, replayed entire baseball games, memorized team rosters, and spent three years trying to remember who had replaced the injured quarterback Y. A. Tittle. (Someone finally told him — Earl Morrall.) Howard Rutledge built five houses in his mind over seven years of captivity, while Danny Glenn's imagined master plan for his house included the location of the joists and studs and the gauge of the electric wire. One night he awakened his cellmate for his opinion on "paneling that family room downstairs."

In the hospital, Cherry's mind allowed him to think of only pleasant things, remembrances he would have never summoned under ordinary conditions. He could see everything that he had ever done and could replay his favorite moments. His father often appeared in these flashbacks, the hot days they worked in the fields. They would not pack a lunch because they'd eat the tomatoes and cucumbers and string beans from the ground, then drink cold water directly from the spring.

He also imagined places he had never been to or things he had never done and made them more real than reality itself. He had never raced a car, but for hours at a time he envisioned himself on the racetrack, behind the wheel, smelling the oil, feeling every turn, maneuvering through the pack, positioning himself for the final lap. He also worked on the car: tuned its engines, adjusted the tires, prepared for the next race. He had never built anything, but now he tried to construct sheds, decks, and porches, each project a puzzle that he alone could solve.

While he thought about Shirley, he also fantasized about having affairs with ordinary women, with "jewels of women," with the most beautiful women in the world — Sophia Loren, Lena Horne, Ava Gardner. He would pick one, think about her all day and night, sleep on it, dream about it, then think of someone else.

He imagined his children: taking them on picnics, riding with them on his motor scooter, and playing with them. His actual dreams, however, were sometimes more realistic: he always went somewhere in these dreams but had to go back. One time he was at home and his daughter was walking down a country path, crying. But he never got to ask her why. He had to return.

In later years, he would consciously think about how to keep his sanity. Elaborate mathematical equations seemed to help. He would calculate how long he had been in captivity — by the year, month, week, day, minute, and second. The next day, he would do it all over again.

One thing he thought about now was Porter. What was he doing? Did he know what had happened to him? Would he be there for him when he returned? And if not, who was going to take care of him?

After twenty-two days, the prison guards came to the hospital, picked him out of his bed, and returned him to the Zoo. No explanation was given. Blindfolded, he careened through the streets of Hanoi, sitting on an inner tube to protect his tailbone. He reached the prison, and his prayers were answered when he was reunited with Halyburton.

Halyburton was elated. "I was worried about you," he said. "What just happened?"

"They kidnapped me out of the hospital," he said.

His shoulder was wrapped well this time, and he had gained some weight, though his tailbone was still protected by only a thin layer of skin. The ride to the prison had put a splinter in the inner tube. Halyburton asked the guards to patch it, which they did. Given all the bicycles in the country, Halyburton reasoned, one thing they could do was repair a rubber tire.

Halyburton and Cherry were then moved to another cell and the following day were moved again, this time to a building known as the Stable. Their new cell had once been a projection room, with a hole in the wall through which the movie was beamed.

Banana trees covered the compound, and many prisoners would mark time by watching the fruit grow, sometimes even getting a few. Halyburton was pleased when he saw guards digging two holes and planting the trees. But they never returned to water them. Day after day the trees baked in the sun, withered, discolored, and died. Their neglect, to Halyburton, seemed to typify the indifference of the Vietnamese to preserving life.

"We're a lot like those trees," Halyburton told Cherry.

No need of Cherry's was too small for Halyburton to fill. One day, for example, he saw that Cherry's fingernails were quite long, so he got clippers from the turnkey and cut the nails on his hands as well as his feet.

One morning, an explosion erupted in the cell.

"Fred, what the hell was that?" Halyburton asked.

Cherry turned to him and said in a small voice, "My tube blew up." The tube that he sat on continually needed to be patched and inflated again, but now it was beyond repair. Halyburton had never seen Cherry look so forlorn, so he grabbed a sweatshirt and placed it, as he had in the past, beneath Cherry in the same position as the tube. Cherry slept on it, and the following morning it was soaked with pus. Halyburton wrung it out and gave it back to him, morning after morning.

Cherry had not washed his hair in more than four months, and when he ran his hand through his thick afro, he pulled out a gob of oil and smeared it on the ground. The oil was set upon by a platoon of ants, who entered the cell from cracks in the wall. Cherry's bunk was right next to these cracks, and the ants soon began to clamber over him in search of his scalp. With his right arm, he tried to swat them off. Halyburton smashed as many as he could and tried to fill the cracks with paper. He made some progress, but the effort was useless as long as Cherry's gooey hair remained a delectable target. Halyburton asked the guard several times if Cherry could wash or take a shower, and he was finally given permission. (Halyburton, who got around the camp more, had periodic showers.)

A water boy nicknamed Johnny Longrifle — he was only fifteen years old and shorter than the rifle he carried — took the prisoners to a shed with a shower room. It had no windows, and when the guard closed the door, they stood in the dark.

Halyburton yelled, Johnny Longrifle opened the door, and Halyburton motioned him to turn on the light. But there was no electrical switch; the bulb was activated by connecting two loose wires that hung near the shower head. The guard, wearing sandals, walked across the wet floor, reached up, and — to Halyburton's amazement — connected the wires. Sparks flew, though no one was shocked.

Cherry was instantly repulsed. Snails were crawling all over the walls and floor, leaving a quarter inch of slime that they were now both standing in. Nonetheless, it was still a shower. Cherry couldn't move his arm, so Halyburton took off Cherry's shorts and turned on the cold water. He soaped his hands and rubbed them into Cherry's hair. At first, nothing much happened, but the lye served as an activating agent, turning the dirt, grease, and water into a paste. He scrubbed it, and gobs of oil and hair and dirt fell out. He rinsed hard as they stood together in paste and slime and soap and cold water, all mixed together like some primordial ooze. But he barely made a dent in the grease on the scalp, which lay a quarter inch thick.

"You won't believe this," Halyburton said. "I'm going to have to wash your hair again."

"You wash it as many times as you need to."

Halyburton washed it again, and again, and yet again, until he could feel Cherry's scalp.

"I think I finally broke through the grease," he announced.

It occurred to Halyburton, at the end, that this was something new for him, scrubbing a black man's hair with his fingers. But Cherry had ceased being black. He was just another American pilot.

Cherry also considered how strange, how unthinkable, this shower would have been in America. Remarkably, the POW camp had made it possible.

9

The Hanoi March

As the summer approached, the Hanoi media increased their attacks on the POWs. They had always been called "criminals," "air pirates," or "mercenaries," but now they were being compared to Nazis and alleged to have committed war crimes, punishable by execution. The Vietnamese hoped to attract attention abroad, and to some extent they did. American antiwar activists in 1966 recounted the "crimes" of U.S. pilots at teach-ins, marches, vigils, and campus demonstrations. The accusations also caught the attention of the Johnson administration, which scrambled to develop a legal brief defending the POWs in case of a trial.

America's bombing campaign escalated significantly on June 29, when 116 U.S. jets dropped nearly a hundred tons of bombs and rockets on oil depots in and around Hanoi and Haiphong. For fifteen minutes, planes could be heard climbing and then screaming down toward their targets. Bomb blasts and antiaircraft fire shook the earth while air raid sirens wailed. Halyburton and Cherry heard them and felt them: plaster fell from the ceiling; cluster bomb pellets raked the roof; the cell vibrated.

They were elated by the attack, believing the war would end only if the enemy was bombed into submission. But the onslaught cost them dearly.

The guards had collected a shirt from each prisoner, then returned it with a three-digit number stenciled across the back. Some numbers ran into the five hundreds, suggesting that Hanoi held more prisoners than it actually did. The motive for the shirts became clear on July 6.

In the late afternoon, the monotony of a hot, humid day at the Zoo was broken by a loud drum. Trucks rumbled into the courtyard; Halyburton and Cherry knew that any change in routine was a bad omen. There was nervous tapping between cells, and the guards on their rounds were more tense than usual. By the early evening, the guards began opening the cells and telling the POWs to put on their shirts.

When a guard reached the cell holding Halyburton and Cherry, he motioned for Halyburton to come alone. Outside, he was blindfolded and led to one of the trucks, which he climbed inside by feeling his way over the tailgate. No one was allowed to talk, but the Americans, using the tap code, communicated by nudging each other with their knees, elbows, or feet, even by coughing. They tried to guess their destination, with the optimists saying that the recent air raid had defeated the Vietnamese and they were now being taken to the airport. The pessimists feared torture sessions in some remote locale. Halyburton leaned toward the latter. In recent interrogations, the Vietnamese had clearly been angry over the raid, threatening to try to execute the Americans. He didn't know where they were going, but he thought nothing good would come of it.

The truck was covered by canvas, so they could hear car horns and bicycle bells, indicating they were in downtown Hanoi. When the truck stopped, they climbed out, had their blindfolds removed, and found themselves in a small park in the middle of a large intersection. There were about sixty Americans, and they searched excitedly for familiar faces. They were soon paired off and handcuffed, and a familiar face — Rabbit's — emerged from a group of officers; he was gripping a battery-operated megaphone.

"You must obey all orders," he yelled. "You must show courtesy. You must be careful. You must bow your heads. Otherwise it could be very dangerous for you.

"You must remember that you are all criminals," Rabbit continued, "and that tonight you are being taken to your public interrogations so that all the world will know your terrible crimes." They were, he said, "about to meet the Vietnamese people." The Americans knew they were in real danger.

They lined up in two columns and separated into four groups. Halyburton and his partner, Air Force Captain Arthur Burer, stood near the front of the second group. Guards in crisp uniforms stood alongside, rifles with fixed bayonets held across their chest. When the prisoners were prodded to march, Air Force Captain Bob Purcell mocked, "Oh, boy, I love a parade!" Rabbit ordered them to show humility by bowing their heads, but Jerry Denton, one of the senior officers, yelled to his troops, "You are Americans! Keep your heads up!"

They walked out of the intersection and headed down a main street. Following them was a long flatbed truck with bright floodlights, whirling cameras, tripods, reflectors, and technicians. The sides of the streets were crammed with people, some standing, others sitting in a grandstand, and most of them yelling. Clusters of foreign journalists, including photographers, were also waiting. Political officers with red scarves escorted the procession, taking it past the Soviet and Chinese embassies to impress officials there with the number of prisoners. Prompted by Rabbit and the loudspeakers, the crowd screamed, "Black criminals!" "Baby killers!" The political officers, meanwhile, kept yelling, "Bow your heads! Bow your heads!"

The intensity of the chanting increased as the throng responded to the officers, some of whom exhorted instructions through bullhorns. Many of the men were dressed alike, in dark blue coveralls, as if they had just left a factory. The women, wearing traditional white blouses and black pants, also raged against the Americans.

Halyburton tried to keep his head up — not in defiance but in hopes of being photographed so that the U.S. military, and ultimately his family, would know he was alive. Every time a light came on, he looked up for a camera, but then a guard would butt him with a rifle or slap him with his hand. Frustration turned to fear as Halyburton realized that the spectators, with overwhelming numbers, were breaking down the protection of the guards. Block after block, the galleries cursed the prisoners, pressing them from either side and throwing bricks, bottles, stones, shoes, and garbage. Scores of Vietnamese darted among the Americans, reaching out, pushing, punching, spitting, screaming. The assailants beat the guards as well. Some Americans took vicious hits — Denton was knocked over with a rock while a famous photograph depicted Air Force Lieutenant Hayden Lockhart holding up a dazed comrade after he had been struck by a bottle. The march lasted for two miles, but the carefully choreographed parade of American humiliation had degenerated into a riot of delirious vengeance.

Halyburton feared that, instead of stones and bottles, he would be attacked with guns and knives. No one knew how the march would end or how the POWs' safety would be secured. Then, suddenly, a file of young men and women in white shirts or blouses, dark pants, and red armbands fanned out on either side of the straggling line. Positioning themselves between the guards and the lunging masses, they clasped hands and formed a new barrier. Then they shouted to the prisoners to run toward a soccer stadium several blocks away.

137

The Americans fled, but the mob pushed through the new protectors, clogged the path, and continued its pounding. Everett Alvarez was pummeled in the back of the head with a blunt object. Navy Lieutenant Gerald Coffee was hit with a fist that split his upper and lower lips. Air Force Sergeant Arthur Cormier's shirt was torn off and his face streaked with blood. Halyburton was also clubbed, first by guards who wanted him to bow his head and then by ruffians, a beating that left him with knots in the back of his head and bruises on his face, shoulders, and arms. He saw a young girl lying on the ground, bloodied, hurt by the mob. He was afraid she would be killed and the Americans would be blamed and tried for murder, but someone picked her up and carried her away.

The guards opened the stadium's rickety metal doors just enough to allow the Americans to slip through. But the crowd continued to surge from behind, forcing the guards to strike out again at their own people. The mob would not be stopped, pushing against the doors and breaking them down. The Americans by now were walking through a narrow tunnel leading to the track, whose secure door could not be breached. They arrived and sat down on the grass, still in a double column. Some believed that they had been taken to the stadium to replay a chapter of Roman history, in which defenseless men would be pitted against wild beasts for the thrills of bloodthirsty fans. But the scene soon had the trappings of a combat evacuation zone, with everyone nursing wounds and the guards taking a head count. The Americans, most of whom were already weak from captivity, were ill prepared to handle such an ordeal, and they felt fortunate that no one had been killed. For the moment they were out of harm's way, but they were not free of derision. A voice over the loudspeaker announced that they had just experienced the wrath of the Vietnamese people.

*　　*　　*

Fred Cherry's night was no better.

After Halyburton was taken away, a guard entered the cell and signaled him to get dressed. He would be leaving again. Cherry wasn't sure where he was going, but he knew the authorities were again interested in his shoulder. Several days earlier, a doctor had examined it and recognized that the bleeding had increased, the stink had worsened, and the infection was raging. Left untreated, he could again be in peril.

On the night of the Hanoi march, Cherry was taken back to the hospital for his third surgery. This time, however, the hostility in the street had permeated the rooms inside. While the previous procedures were impaired by the staff's poor medical skills, Cherry believed the doctors had tried to heal him and to minimize his discomfort. That was no longer true.

Cherry recognized the doctor, a short, rotund man who in previous operations had offered comforting words. Tonight he was silent and dour. He used a pointed surgical instrument to tap the shoulder, but he pressed too hard. The tip punctured the brittle skin and slid right through. Meanwhile, a nurse prepared the injection that would knock him out. She inserted the needle and began to press, but the doctor stopped her. She withdrew the needle, and Cherry realized he would be awake during this operation.

The doctor placed a small white sheet over Cherry's face, neck, and good shoulder, then he took a scalpel and cut away the dead flesh of the crippled shoulder, literally scraping the infection off the bone. Perspiration poured out of Cherry's body; it was, he later said, "the worst straight pain I've ever felt," an act of sadism. He couldn't believe that a man who was supposed to heal would hurt him so badly. But he refused to cry out. He heard the doctors and nurses talking; it seemed they were puzzled that their patient remained conscious but didn't scream. Cherry believed it was a test of wills, and they were not going to break his. He was not going to plead for his life. Gritting his teeth until he thought they would break or fuse together, he made sure he was smiling whenever the sheet was raised from his face. For at least an hour, he kept saying to himself that he could make it. When they finished scraping and washing the bone, bandaging the shoulder, and wrapping the arm, the same grin appeared on his face. While the operation was his most excruciating one, it finally arrested the infection.

Cherry was taken back to his cell, where Halyburton had just returned from his own harrowing night. He noticed that Cherry's sling was already a bright red.

"Fred, what in the world did they do to you?"

"Oh, Haly," was all he could say.

Halyburton carried him to his bunk as the blood spilled out of Cherry's shoulder, running all the way to his feet and drenching everything around him, including Halyburton.

"I can't believe they took you on the march as well," he said.

"I haven't been on a march," Cherry said. "Where have you been?"

"They marched us through the streets of Hanoi and they kicked the shit out of us."

Cherry described his ordeal.

"It looks like we've both been through it," Halyburton said.

"Do you know what you look like?" Cherry asked.

"No, but I know what I feel like."

"Well, you look pretty bad too."

Outside the cell the brutality continued, as Americans were tied to trees in the courtyard, locked in small black cells, and berated during interrogations. The following morning, Bob Purcell tapped to another prisoner, "I hope this doesn't sound too presumptuous, but after last night I think I almost — up to a point — know how Christ must have felt."

The march itself backfired against the North Vietnamese. They had hoped to photograph and film humbled Americans before a victimized but orderly population; they instead captured manacled prisoners being terrorized by locals. Rather than generating international sympathy for their claim of war crimes, they provoked anger even among American doves who had previously defended Hanoi. The Johnson administration vowed to use overwhelming force if the POWs were tried as war criminals, and, prodded by public outrage, the administration became far more sensitive to the captives' plight. The government, never having formally declared war in the first place, had been reluctant to say "prisoners of war," but a new Committee on Prisoner Matters was given the resources to contact other governments, solicit support from international aid agencies, and even initiate conversations with Hanoi about the captives' welfare.

The days after the march also marked a new era of extreme hardship for the prisoners. The Vietnamese, still outraged over the recent bombing but now embarrassed as well, vented their frustrations at the prisoners. His interrogators threatened Halyburton that he would be tried and executed as a war criminal, and McGoo struck him with a fist. These tactics had been used before, but typically in a haphazard or bureaucratic manner. Now his captors were genuinely angry, eager to harm the Americans.

Halyburton and Cherry were unaware of the international reaction to the march. All they knew was that the Vietnamese were preparing for another purge. Halyburton continued to change Cherry's bandages, wash the soaked sweatshirt every morning, and knock the ants away. In this new climate of oppression, both men took comfort from their having each other.

The prison officials, however, finally realized that their scheme had failed, that they had not put Halyburton in a "worse place" when they locked him up with their only black captive. By now they knew that Halyburton had not just endured his roommate but saved him. On the night of July 11, a guard walked in and motioned to Halyburton to roll up his clothes. As always, he began to roll up Cherry's belongings first, but the guard said, "No, just you."

Both men knew what that meant: if you leave without your bedroll, you're coming back; if you take it, you're gone for good. Their time together was over.

"Fred, I'm sure you're going to be okay, but it looks like this is it," Halyburton said.

Cherry's eyes welled up as Halyburton gathered his clothes. Cherry leaned forward, pulled something from behind his back, and stood up. "Haly, you forgot your sweatshirt."

"You keep it."

The two men hugged, cried, and said good-bye. Halyburton walked out, leaving his roommate standing in the middle of the cell. Cherry later said, "That was the most lonesome night I ever spent in my life."

Their time together, seven and a half months, would represent less than ten percent of their time as prisoners, yet it was the turning point of each man's captivity.

They had been forced to live in a very different world than they had ever known. That world was harsh, but it also had a kind of pureness and clarity of purpose. Survival transcended all other concerns, and traditional sources of tension — race, service, rank, family background — were replaced by the bonds of compassion and sacrifice. It would not be the last time that Halyburton and Cherry would create a rarefied existence in their prison cell.

Every POW confronts the same question — Do I want to live or die? — and everyone answers it in deed or thought. The closer he is to death, however, the harder it is to maintain his will to survive.

But as a former Air Force psychologist, Ludwig Spolyar, wrote in 1973, "If he does decide to survive, everything he does from then on is dependent on the need to survive." At that point the prisoner can accept his suffering as "necessary for the fulfillment of future freedom," and he can "endure the suffering, which now has meaning to him."

The symbiotic friendship between Cherry and Halyburton had layers of meaning. Cherry would have died without his roommate — but in surviving he also rescued Halyburton from his despair. Each man inspired the other and helped make that elemental decision, to live or die, easier.

Cherry knew that Halyburton could have done less, or nothing, and no one would have held him accountable. That Halyburton was a white man from the South deepened his gratitude. Cherry's self-sufficiency had always helped him get along with whites. He made few demands and took pains not to discomfit others. But as a prisoner, he was in perpetual need and was at times an invalid. What's more, their relationship completely reversed the social order in which each had been raised: the white man was the ultimate servant to the black man, feeding him, exercising him, bathing him, sweeping the room, and more. It was not a true role reversal: in the segregated South, blacks had to perform those jobs to survive, whereas Halyburton did them voluntarily, without compensation or acclaim. Cherry now realized that some Southern whites were capable of more humanity and generosity than he had thought possible. "I didn't expect Haly to do all that," Cherry said years later. "He was white and he was from the South, but he taught me that you can grow up in that environment and separate the good from the bad and the right from the wrong. He was one who did that."

Cherry's impact on Halyburton was less visible but equally vital. All Halyburton's energy had been focused on his own plight, his own discomfort, and he had reached an emotional nadir before meeting Cherry. Their friendship renewed his spirits and motivated him to find meaning in his captivity.

"Caring for Fred," he later wrote, "was the first opportunity I had to turn from my own concerns to another's. I realized how trivial mine were by comparison and how he bore his pain and suffering with such dignity — never complaining, never cursing his fate, never giving up. The task of caring for him gave a definite purpose to my immediate existence, and it was a task that I gratefully took up. In the process, I received much more from him than I was able to give."

The experience forced Halyburton to confront his own bigotry. Cherry was certainly not limited, as Halyburton had always thought of blacks. Indeed, Cherry's life was testimony to unlimited imagination, ambition, and skill; he mastered the intricacies of combat flying while displaying courage, resilience, and patriotism. Halyburton had never considered such a possibility, that blacks were his equal. Yet he had to acknowledge that Fred Cherry was his superior — in rank and in reality.

"I was in awe of him," he said, "and I had learned to love him."

10

The Home Front

The families of the early POWs were racked by uncertainty, not knowing the condition of their loved one, how he was being treated, or whether he would return. One wife said her husband was "with the living dead." Coping took many different forms, but the wives of Halyburton and Cherry responded in extreme ways — for good and bad.

In October of 1965, Marty Halyburton was visiting an aunt in Atlanta when she heard a car pull up and several doors slam. She looked out the window and saw an official Navy car; approaching the house were two men in uniform and a chaplain. They had found Marty by speaking with her relatives in Florida, whose house she had just left. Marty knew there was only one reason that a chaplain would visit a serviceman's wife at home: something had happened to Porter.

"The plane was shot down," an officer told her. Halyburton's jet hit a mountain, though his status — killed in action or missing in action — had not been determined. He held out little hope. "There was no parachute sighting," he said. He briefly described the mission, showed Marty a map of North Vietnam, and pointed to the site of the crash. He promised to return as soon as possible with definitive information on Porter's status. Words of comfort were probably offered, but Marty did not remember them.

The following day the men returned, and the officer provided a few more details. A Navy helicopter had flown over the crash site but had failed to detect any radio signals that would have come from the emergency beeper. He repeated that no one had seen a chute. "We know he's dead," he told her.

For the next several days, Marty looked at her baby and realized Dabney would never know her father. But it was a thought she could not yet absorb. She was numb and terrified. Her own mother had died several years earlier; an only child, she was not close to her gambling father. And now, at twenty-three, she was also a widow and a single parent.

A friend drove her and Dabney to Davidson for a memorial service. In a town of a thousand people, the death of one young man was a huge loss. Virtually everyone knew Porter and his family; his grandparents had lived there for more than forty years. Flags were at half mast at the post office, the town hall, and Davidson College. The football team canceled practice on October 21. Two days later, before its homecoming game, the squad paid tribute to Halyburton with a moment of silence. Attending the game was Halyburton's mother, Katharine, who worked in the school's public affairs department.

A gravestone for Porter, marking the dates of his birth and death, was placed in a family plot.

Katharine received dozens of condolence letters, including one from Jimmy Woods, the son of a Davidson doctor who had grown up with Porter. He was now an Army captain in the 35th Infantry Regiment. "I know how much he loved his family and his country," Woods wrote. ". . . he died a hero." A few months later, on February 6, 1967, Jimmy Woods was killed by friendly fire in Vietnam's Pleiku Province. He was twenty-five. Many Davidsonians agonized that this war had claimed two young men from so small a town.

The memorial service for Halyburton was held on October 21 at St. Albans Episcopal Church, a low brick structure that blended inconspicuously with a residential neighborhood. Parking signs were posted to direct traffic. A photographer from the college strolled around taking pictures. More than 150 people attended, well more than the church could seat. Speakers were set up outside for the overflow. It was a sunny afternoon, but the church was dim. An organist played "The Strife Is O'er" and the Navy hymn, "Eternal Father, Strong to Save." Visitors signed a guest book; on the page for Final Resting Place was written "North Vietnam." Attendees also received a booklet with a collection of Halyburton's poems, mostly innocent riffs with such titles as "The Cruel Sea," "An Autumn Fire," and "Katharine," a tribute to his mother. No one disputed Katharine Halyburton's dedication, which said the poems "reveal a depth of soul, a maturing understanding, and a heart and mind sensitive to beauty." The booklet was titled "Thoughts Before 21," the poems having been written by the teenage Porter.

Both Katharine and Marty were escorted by Navy officers and sat in the front row. An honor guard entered the church, folded two American flags, and handed them to the Navy escorts, who presented them to mother and wife. A friend held Dabney, whose soft cries were heard between the prayers and the hymns. Katharine wore her son's pilot wings on her lapel and was remarkably composed. She did not cry during the service; at times, even a trace of a smile appeared — it seemed that she was trying to comfort others. She found less conspicuous ways to express her emotion. Sitting next to her was Bill Thompson, a friend of Porter's, and she squeezed his hand when a poignant moment passed.

While the church was Episcopalian, the ceremony was Presbyterian, and it reflected an ethos of restraint and sobriety. The purpose of the service was more to worship God than to praise the dead. The chaplain, Will Terry, knew the Halyburtons well, and this service was one of his most challenging. Porter's death had spurred the antiwar sentiment — Davidson was, after all, a college town — and the unrest heightened everyone's emotions. Terry rewrote his "meditation" on Halyburton four or five times.

Standing on the altar, he spoke firmly, noting that a "memorial service" was a fitting ceremony because "a memorial is something that keeps remembrance alive." Alive first, he said, "is the memory of Porter Halyburton as a friend, a husband, and a son who brought joy rather than pain," and whose memories of "tenderness, devotion, and considerateness fill our hearts with gratitude toward God, who is the origin and author of such virtues." But these memories, he said, are also jarred by tragedy — not of death, for we all must die, but from "the kind of death that comes from the insanity of war . . . from the collective selfishness and callousness of man . . . and in this sense we are very much participants in his death." But there is yet another memory, Terry said, which he called "the recollection of victory." It is the memory of the resurrection "that allows us to look death in the face . . . This is the certitude that God has blessed Porter with. It is the victory of God, and it is our victory as His children. This does not keep our grief from being grief. The separation is real, but the reunion is also real, and this is what saves us from despair . . . Our memories end not in grimness, but in the quiet joy of genuine gratitude."

After the service, friends and relatives gathered at Katharine's house, where Porter had grown up, and was soon filled with dishes of spaghetti, baked apples, date bars, and layer cake. Katharine retained her poise, expressing pride in her son, showing little grief, and assuming a fatalistic air. As she told her friend Nancy Blackwell, "If this is the way it was meant to be, then this is the way it's meant to be." Her parents were alive but ailing, and she had to be strong for them. But her steely pose was deceiving. As a public relations professional, she knew the importance of image — her bold scarves and costume jewelry had created a cosmopolitan persona — and she did not want to show her vulnerability. Her friends, however, knew better, and understood how much her own life was vested in her son's. Divorced and remaining single, she had quit her promising newspaper career so she could tend to her parents in Davidson. What she'd had was Porter, and now he was gone. As Erskine Sproul, a photographer who worked with Katharine, said, "She was able to continue her job stoically, but her eyes were always red."

On the night she learned that her husband had been shot down, Shirley Cherry gathered her children and told them what had happened. Her second child, Fred Jr., who was ten years old, didn't understand. "What does that mean?" he asked.

It meant leaving Japan and returning to the United States, which surprised some of Cherry's friends. Other wives in such circumstances stayed in Japan, where information was more accessible. Shirley and the children initially moved in with Fred's oldest sister, Beulah, and her husband, Melvin, in Suffolk, Virginia, for they could accommodate Fred's family. On Sundays the children would visit their grandmother in the same farmhouse where their father had once lived.

The arrangement, however, was difficult for the youngsters, who did not respond well to Beulah's discipline. Adjusting to public school was even more difficult. While military schools were rigid, the Suffolk schools were lax, with children cursing, fighting, and ignoring their teachers. "It was a culture shock," Fred Jr. said. "I just wanted to get home, crawl under the bed, and stay there."

Before long, Shirley and the children moved twelve miles east to Portsmouth, which coincided with alarming news for the chil-

149

dren: their dad had not survived. "I always thought he was dead," said Cynthia, who was six when her father was shot down. "That's what Mom always said."*

In fact, there is no evidence that Cherry's status had changed. In a letter to his mother, Leolia, five months after the shootdown, the Air Force said it knew that he had "ejected, deployed a good parachute, and was observed to land on a small hill in a slight ravine." He signaled with his radio beeper, but voice contact was never made. Whether he survived was uncertain, the letter said, but "it is reasonable to assume that he may have been taken captive, and is being detained."

In the following three years, the Air Force continued to write to Leolia and Beulah without indicating any change in Cherry's status. The Air Force was proceeding as if he were alive. In 1968 Leolia was told that her son had been promoted from major to lieutenant colonel.

By then, Shirley had severed her ties with Fred's family. When Beulah tried to visit the children, Shirley would not let her see them, so Beulah would park her car and watch the kids play on a basketball court.

Shirley continued to receive money from the Air Force — according to Cynthia, $1,432 a month. They moved from an apartment in Portsmouth to a single-family house, and Shirley began dating. When Sam Morgan, an Air Force pilot who had known the Cherrys in Japan, visited Shirley in Virginia, he was struck by how much she had changed. She had lost weight, bought new clothes, and changed her hair style. "When I knew her in Japan, she was the classic Air Force wife, but when I saw her in Virginia, she had decided that Fred was gone and she was a completely different

---

*Fred Jr. and Debbie confirm that Shirley told them their father was dead, but Don said his mother told him he was missing in action.

person," he said. There was another man in the house, he added, "and she was not happy to see me."

To be sure, Shirley was hardly the only POW wife to start a relationship with another man. As one wife told the *New York Times* in 1972, "I don't want to live as if I were dead." Another woman said, "A lot decided to stay faithful until they met a man who was attractive." The stress on the wives caused many marital breakups. At the end of the war, the Pentagon said that 39 of the 420 returning married prisoners, or 9.2 percent, "have either gotten or were getting divorces." Ten years later, *U.S. News & World Report* said that at least ninety couples, or 21 percent, had divorced. But those numbers represent all married prisoners, including those caught toward the end of the war; the rate for the early shootdowns was higher. Few breakups, however, were so one-sidedly bitter and gratuitously hateful as the Cherrys'.

According to Cynthia, her mother seemed to take pride in telling others what her final words to Fred were: "She said the last thing she ever told him was, 'I hope you get shot down.' And then he got shot down."

Fred's mother suffered a stroke the same year he was captured. Until then, the seventy-eight-year-old had remained spry, still raising chickens, canning fruits, singing out the Lord's name, and threatening malefactors with her .38 revolver. But the stroke slowed her down, and the news of her son was almost more than she could bear. "It was like someone had knocked the wind out of her," said Evelyn Brown, a granddaughter. "She didn't display her emotions and she didn't cry, but you knew from her expression that she was torn up inside."

But as long as Fred was missing in action, she could hope he was alive, and that hope sustained her. She was more confident than anyone else in the family. Each day, surrounded by photographs of Fred in his uniform, she sat in her chair with her head down. She would look up if someone spoke to her; otherwise, people assumed she was praying. Before meals she would say a blessing: "Lord, please take care of my baby, and let me live to see him come home." She did not allow herself to believe otherwise. "He'll be back," she said. "I may not be alive to see it, but he'll be back."

Beulah was now the family matriarch and the principal contact for the Air Force. For three years there was little news, and any information was either misleading or unnerving. In March 1967, the Air Force wrote to Beulah that an American rabbi and minister had visited two unnamed American POWs in North Vietnam, and the pilots "were said to be in reasonably good spirits and receiving adequate medical treatment." But later that month, the Air Force wrote that "increased pressure is quite possibly being brought to bear on our personnel being held captive . . . to force them to make unfavorable statements against their country." There was no indication about what type of pressure or how it was applied — nor, for that matter, any new evidence on whether Fred was dead or alive.

The American National Red Cross took packages to the POWs, so Beulah sent some to her brother, hoping that he was in fact a prisoner. But she was less confident than her mother. Typically buoyant and chatty, Beulah was now often quiet, detached. Friends and relatives picked times to talk with her about Fred, recounting the good times and laughing at his antics.

There were, most famously, his low flights over the community. When he was stationed at Dover, he loved to take detours and buzz the houses in his old neighborhood. Sometimes he would alert his family and friends to "the show," and everyone would stand out in the yard and wait until someone would finally hear a distant rumble. It got louder and louder and soon muffled the screams of the children. Necks would strain, and then this streak of metal would roar right above the treetops, smoke blowing out its rear and spectators gleefully diving for cover. The plane would climb, swoop, twirl, and dive, then peel away and disappear.

Fred loved to impress Beulah. One time, when he was about to land at nearby Langley Air Force Base, he instructed the radio control tower to call his sister.

"Mrs. Watts, your brother is coming in."

"When is he coming?" she asked.

Right on cue, Fred buzzed her roof and rattled the house. "He's here!" she cried, throwing the telephone in the air and running to the window to catch a glimpse.

On another occasion, Beulah took Fred to Langley and was allowed to go up to the control tower to watch her brother in an F-105. At dusk, with the afterburners glowing, Fred waved as he taxied by. He was going to Mobile, Alabama, which at normal flight speed would take an hour and a half. But that evening he flew at supersonic speed, or about 900 mph, which is forbidden across the continental United States, but it allowed Cherry to reach Mobile in an hour. Landing, he jumped out of the plane, raced to a telephone, and called Beulah, who had just returned from the air base.

"What's wrong?" she asked. "Are you back at Langley?"

"No, I'm in Mobile, Alabama," he said proudly.

She was stunned but pleased that no barrier — sound or racial — could impede her baby brother.

These memories, shared with friends and family, made Beulah smile and would lift her spirits, but then she would grow silent, the pain returning to her eyes. Everyone knew it was time to stop talking.

11

"Unspeakable Agony of the Soul"

Fred Cherry and Porter Halyburton were separated for the rest of their captivity, but their time together had prepared them for the hardships to come. As the enemy's purges intensified, both men suffered the worst treatment of their imprisonment. But their friendship remained a source of inspiration, giving each man additional incentive to resist and endure. Neither Cherry nor Halyburton wanted to let the other down, to negate what had passed between them. As Cherry said, "After all Haly had done for me, I wasn't about to disappoint him."

The treatment of Cherry was always complicated by race. The Vietnamese thought they could demoralize and divide the Americans — and minorities were the ideal target. If African Americans were denied civil rights in their own country, why should they fight and die for that country in Vietnam?

This message was delivered in many ways. In 1965 Hanoi played a tape of Clarence Adams, a black former Army sergeant from Memphis, ridiculing the United States. Adams had been taken prisoner during the Korean War and refused repatriation because of racism back home. In 1966 the civil rights activist Diane Nash Bevel visited Hanoi and saw a movie about Ho Chi Minh, which said he had visited Harlem in his youth and had "resented the exploitation of Negroes in the United States." Bevel, in a broadcast to her "black brothers" in South Vietnam, said, "The Vietnam War is a colonialist war. If you fight in it, you are fighting Asian brothers who are determined to prevent their country from becoming owned and managed by racist capitalist white men."

The following year, the Viet Cong released two black POWs, Army Sergeants Edward Johnson and James Jackson, along with a statement expressing "solidarity and support for the just struggle of the U.S. Negroes . . . for basic national rights." The Viet Cong also dropped propaganda leaflets on National Route 13, a major highway in Bing Long Province, urging black soldiers to join the enemy.

The Vietnamese correctly anticipated that black Americans would turn against the war much earlier than whites; opposition stemmed not just from militant leaders like Stokely Carmichael and Malcolm X but from mainstream black men as well. The most prominent voice belonged to Martin Luther King, Jr., who concluded that Vietnam was needlessly destroying lives, white and black, American and Vietnamese, while undermining domestic programs for the poor. He was also angry that his government would sacrifice blood and treasure to protect the rights of the Vietnamese but not do the same for its own citizens. "We were taking the black young men who had been crippled by our society," he said in a speech in New York on April 4, 1967,

> and sending them eight thousand miles away to guarantee liberties in Southeast Asia which they had not found in southwest Georgia and East Harlem. So we have been repeatedly faced with the cruel irony of watching Negro and white boys on TV screens as they kill and die together for a nation that has been unable to seat them together in the same schools. So we watch them in brutal solidarity burning the huts of a poor village, but we realize that they would never live on the same block in Detroit.

There is little evidence, however, that the antiwar statements from civil rights leaders or other racially motivated propaganda affected significant numbers of African American soldiers. While they were conscious of race, only rarely did that awareness translate into sympathy for an enemy that was otherwise trying to kill them.

That was why the North Vietnamese believed that Fred Cherry, a decorated combat pilot, a career serviceman, and the highest-ranking black POW, was so valuable — and why they used every means possible, including torture, to extract a statement from him. His credentials would carry weight with African Americans in the battlefield as well as on the streets back home. If Cherry denounced his government for pursuing a morally bankrupt foreign policy, if he urged black soldiers to fight for economic and social justice instead, or if he instructed African Americans in the United States to reject the war, Hanoi would claim its greatest propaganda victory. It would release him with great fanfare, broadcast his views on radio and television, and make him a hero in developing and socialist countries. If Martin Luther King, Jr., opposed the war, then why wouldn't Fred Cherry?

But Cherry was a military officer who believed in his country and its war aims in Vietnam. Even in the face of torture, he would not repudiate America. Doing so would affirm the negative stereotypes, including lack of courage and patriotism, that had long tarnished African Americans in the armed services.

"I know how some white Americans feel about blacks," he said years later. "If I do one little thing wrong, they're going to multiply that and you'll hear the same old thing: blacks aren't capable of doing this or that, they can't stand up under pressure, and they're not loyal to their country, which is the damnedest thing I've ever heard. Well, it wasn't going to happen on my watch. I was fighting for twenty-three million black folks. That was my battle."

*   *   *

Halyburton's battle was more traditional but no less powerful: he was fighting to uphold the Code of Conduct for U.S. armed forces.

The code itself was written by the Department of Defense in 1955 amid rising concerns, even hysteria, over the conduct of the American POWs in Korea. Studies described massive collaboration by American prisoners with the enemy, their failure to resist Communist propaganda, indoctrination, and even brainwashing. In short, the American fighting man, when captured, was soft — some scholars blamed domineering mothers — and the military had to move quickly to fortify its troops.

These claims were at times exaggerated and often contemptible, destroying the reputations of so many servicemen who endured disease, hypothermia, starvation, and torture. But during the McCarthy era, virtually any hint of weakness against communism escalated into a national crisis, particularly if centered on the military. Thus, the armed services organized programs to prepare their men for captivity in the next war, and standards of behavior were set forth in the Code of Conduct. All servicemen — and women — learned the code in the classroom, and its lessons were reinforced by posters on bulletin boards and office walls.

Its most controversial statement centered on what information a captured serviceman could disclose to the enemy. According to the code, he is "bound to give only name, rank, service number, and date of birth." It was, to some scholars, a guideline for behavior, not an absolute dictum. But the Navy, to Halyburton's dismay, interpreted it literally.

A year before he was shot down, Halyburton was sent to survival school, another by-product of the Korean experience. In a simulated war game, he was set loose in the woods, captured by the "enemy," placed in mock prisons, and even abused. He was shoved into a footlocker, forced to kneel on pencils (like a knife in the knees), slapped around, berated, and humiliated. It was all designed to break him down, to induce him to disclose any information besides the "big four" of the code. In one training scenario, Halyburton was told he had a chest wound that needed a blood transfusion.

"What type of blood do you need?" his captor asked.

"A positive," Halyburton said.

Wrong! Halyburton was reprimanded for revealing his blood type; it was outside the code.

Halyburton thought this hard-line approach was silly. His blood type, for example, was no secret — it was already noted on his dog tags, his military identification card, and his Geneva Convention card. He agreed that no prisoner should make statements disloyal to his country or reveal relevant military information. But why shouldn't a prisoner disclose, say, his medical needs or agree to write a letter home if its contents conveyed information about his status? Halyburton believed such communications were acceptable. The real issue was how strictly should one follow the code in light of threats or torture? How much pain should one endure before cooperating with the enemy, and would such compromises constitute treason? What was more important — honor to country or to life itself?

For Halyburton as well as Cherry, these questions were no longer hypothetical.

Torture had previously been applied to punish only the most recalcitrant prisoners; but now, for at least a year beginning in the summer of 1966, it would be used indiscriminately. The Vietnamese, embarrassed by the Hanoi march but still determined to press criminal charges, wanted to wring out "confessions" that would confirm the aviators' illegal actions while acknowledging their captors' humane treatment. Even if the Vietnamese recognized the obvious — that coerced statements would be ignored by any legitimate court and discounted by world opinion — they still had propaganda value for the Communists' efforts to rally their own people to their revolutionary cause.

Immediately after the Hanoi march, North Vietnam broadcast over the radio several "confessions" or "apologies" allegedly issued by Americans. In one, Navy Commander James Mulligan was quoted as saying:

> This war in Vietnam has no appeal for me for it was an unjust war against a people who never did anything to the detriment of the U.S. interests. My military obligation forces me to participate in this war; many other military men share this same attitude . . . For my own crime I beg your forgiveness and request that you treat me humanely and allow me to have some part in ending this dirty war waged by our government.

The Communists would go on to seek similar statements from almost all POWs.

After Halyburton departed, Cherry remained at the Zoo and was eventually moved into a building called the Barn. His roommates were two experienced Air Force majors: Lawrence Guarino, a firebrand from New Jersey who was forty-three when he was shot down in 1965 and who spearheaded much of the prisoner resistance with his hard-line opposition; and Ronald Byrne, a Korean War veteran from New York who combined a strong religious faith with a sense of duty. The three men lived together during the early days of another campwide purge, fueled by U.S. bombings. The crackdown included twice-daily room searches, beatings over such infractions as improperly folding one's bedding, and bare-knuckle quizzes.

Cherry did not appear to be in bad shape until he took off his shirt and shoulder bandage. Then he revealed a quarter-size hole oozing fluid — the color of "pale strawberries," Byrne said — and, beneath that, muscle tissue. According to Guarino, the shoulder "had no muscle tissue left and looked like a wire clothes hanger." Cherry explained the history of the injury; Byrne, incredulous, asked how he had survived.

With his typical stoicism, Cherry initially said, "This is the way things are, and that's the way things have to be." But he also talked about Halyburton's contributions in sustaining him throughout his near-death drama — demanding medical help, dragging him around the floor to keep him alive, feeding and bathing him. "I was down to eighty-five pounds," Cherry said. "If it wasn't for Porter, I wouldn't be here."

On August 15, Guarino was taken in for interrogation, and when he returned he said, "We're going into torture."

Cherry was then taken into interrogation. "You have a bad attitude, and you disobey camp regulations," Dum Dum said. "You communicate with other criminals. You must be punished. You must have 'iron discipline.'"

He was returned to a small building called the Gym, where Guarino and Byrne had already been taken, and the guards dragged in manacles and leg irons for each man. When they tried to clasp Cherry's wrists behind his back, his left arm could not be twisted up and back. He screamed while Byrne yelled, "No! No! No! You'll break it!" The guards, agreeing, used nylon rope instead, still tying his wrists behind his back but without twisting them. Because the rope could be untied by a cellmate, Cherry was returned to his previous cell, where the leg irons were slapped on.

He was alone, released twice a day to eat and wash but otherwise tied up and locked in. When he refused to confess his crimes or condemn his government, he suffered the "fan belt treatment," in which a guard beat him with bamboo or strips of rubber, raising welts on his back.

He had no place to go and no one to speak with. He wouldn't cooperate with the Vietnamese but treated them as he did the bigots at home. He was firm but not antagonistic, using humility as a defensive maneuver. When an interrogator demanded that he write something, Cherry said, "I wouldn't ask you to do this. Why are you asking me?" He told them they were both military men and that this was not the way soldiers treat each other. His strategy was to avoid writing that first word, which could lead to a stream of regrettable statements. The Vietnamese, of course, did all they could to prompt that first word by asking innocuous questions.

"What do you do for Easter?" the interrogator asked.

Cherry picked up the pen and drew eggs and children, but he never wrote any words.

Cherry used his race to his advantage, playing ignorant, agreeing that black Americans received poor educations, and saying that he simply didn't know enough to answer questions about his government or the war. "I've been out of the limelight of things," he said.

For ninety-three days, Cherry mostly sat on his bed board, chained and tied like an animal. In his frail condition, any mistreatment constituted torture. Guarino wrote in his official report: "Taking advantage of his disabilities . . . the Vietnamese put Major Cherry in leg irons and then tied his arms together. This comparatively low-key method of coercion nearly cost Major Cherry his life because it aggravated his wound considerably."

During the ordeal, American prisoners in an adjacent room heard soft bumps through the wall and recognized the tap code. The noise came from Cherry's cell, but how could he tap with his legs and arms bound? They figured it out: Cherry was knocking his head against the cement wall, and his message was not idle chatter but important information about enemy tactics. As Air Force Captain Bob Lilly, in a nearby cell, later wrote: Cherry's words described "what the V wanted and how he was avoiding giving it to them. It was important to the rest of us to know what was working and what was not. Any insight into what they were after helped prepare a defense."

Lilly wasn't surprised. When Cherry roomed with Halyburton, Lilly had heard Fred struggling to breathe while Porter tapped updates on his condition. But now Cherry's resilience and resourcefulness — his ability to endure unimaginable pain, defy his captors, and still assist fellow prisoners — took on a mythic, superhuman quality.

"The V were demanding all would write a biography which would lead to confessions of crimes," Lilly recalled. "All eventually wrote something, except Fred. To my knowledge, he never gave in and wrote what the V wanted. [When] the V finally realized he was going to die before he would write, they let him off the hook. This is the only time that I know of that anyone outlasted the V . . . Fred did not comply and he did not die or lose his mind. In so doing, he became a legend to the POWs."

After leaving Cherry, Halyburton was blindfolded and driven to a mountainous region near the town of Xom Ap Lo, thirty-five miles west of Hanoi. He thought he had already seen the worst Vietnamese prisons, but he was about to discover a new standard of primitive living. The compound was configured like a tic-tac-toe grid, with high concrete walls enclosing nine different brick buildings, each with four tiny cells, about eight feet square. A built-in bed board and a narrow foxhole, dug out in case of an air raid, left little open floor space. There was no electricity, no plumbing, no place to walk around, and no medic. Meals consisted of rice, the summer heat turned the cells into steam baths, and the prison's remote location attracted the surliest guards, who could punish inmates without restraint. The Americans had a fitting name for this slice of hell: the Briarpatch.

Halyburton was often handcuffed, given just enough slack to relieve himself. The dark cell obscured alien matter in his food, including the cockroach he once bit. The foxhole bred mosquitoes. At one point he developed a bad cold and used a small towel as a handkerchief. Unable to wash it, he hung up the mucus-covered cloth at night so the ants would clean it off.

Hygiene, however, was the least of his problems. The Vietnamese introduced a Make Your Choice program; the Americans could "choose" cooperation or defiance. One would lead to good treatment and possibly early release; the other, to torture and possibly death. This style of interrogation had already been used on Halyburton; his answers would determine if he were sent to "a better place or worse place." The main differences now were that the Vietnamese wanted damning statements that could be used against America in a public forum, and they were willing to inflict far greater punishment to extort them.

The Briarpatch commander, whose accent earned him the nickname Frenchy, carried out the program with gusto. He was assisted by certain guards, such as McGoo and Slugger, who would force prisoners to run barefoot and blindfolded through the compound or would drag them with a noose around their neck. Even a "chow girl" nicknamed Flower demanded bows from the Americans. But the most notorious interrogator was Bug, also called Mr. Blue for the color of his uniform. Mr. Blue was short and fat, his wandering right eye — clouded white by a cataract — evoking scorn and terror. He was also emotionally combustible, constantly jabbing his right index finger and harping, "You have murder my mother."

He began his torture sessions with Air Force Captain Paul Kari and proceeded through the ranks. J. B. McKamey. Everett Alvarez. Tom Barrett. Scotty Morgan. All were brutalized, their screams heard day and night. Halyburton heard the cries, and when he was called, he was terrified.

Sitting behind a felt-covered table, Bug was waiting for him when he entered. The American sat on a stool as two guards stood nearby. Bug placed a piece of paper in front of Halyburton and demanded that he write his confession of war crimes as well as his biography. "It's time for you to choose," he said.

Halyburton had been given instructions on how to respond to such a threat. Lieutenant Colonel Risner had issued an order that the Americans should resist to the utmost, give as little as possible, and then recover to resist again — a variation on the Code that acknowledged the Americans' dire circumstances. Halyburton was determined to follow these instructions, and he prayed for the strength to endure.

His training, his background, and his sense of honor gave him his own code of conduct: he would rather die than give in. He believed that resistance, however painful, was fundamental to the morale and unity of the group, and that failure to resist was equivalent to collaboration with the enemy. If you didn't accept torture, you were a disgrace to all the POWs.

He also remembered his experience with Fred Cherry. He had seen how much Fred had suffered without complaint, and he would never want to fall short of that example or feel that he had given less than his friend.

"It's time for you to choose," Bug repeated.

"I'm not going to work against my government," Halyburton said.

Bug made his demand again and received the same response. His anger rose and one eye went "buggy," moving out of sync with the other. Sometimes interrogators would pretend to be outraged, but not Bug. Halyburton knew that he may have been a psychopath, but at least he was sincere. "Life will be very different for you now," Bug said. He signaled to the guards and left the room.

The guards knocked Halyburton off the stool and, sitting on

the ground, forced him to lean forward with his legs straight out. They pulled his arms behind his back, tied his wrists together with rope, and lifted them straight up. At first he tried to deflect the pain by recalling soothing moments, like running on the beach or dancing with Marty. But the stretching felt as if his body were being torn apart. Pain shot through his shoulders, elbows, wrists, and back. At one point, he could see his fingertips over his head. He screamed.

With his arms still behind his back, the guards slapped a pair of "ratchet cuffs" on his wrists and slid them down his forearms. Normally, the short chain linking the cuffs would fall between the wrists, but the guards attached the cuffs so that the chain wrapped around the outside, a technique designed to intensify the pain by putting extreme pressure on his twisted forearms. Then the guards placed him in leg irons and draped a rope around his neck, which was used to tie his head to the irons. Pulling the rope forced Halyburton to bend even farther, inflicting greater stress on his body and further tightening the cuffs around his forearms. His body was being torn apart, the cuffs seeming to cut through to the bone. The guards then released the first rope around his wrists, but the movement of his contorted arms returning to their normal position caused the rigid cuffs to twist more deeply into his skin and bone. He groaned, screamed, saw white. He prayed and held out for several hours, his assailants manipulating the ropes and cuffs, the pain coming in waves and sunbursts. It was the torturer's ultimate cunning: the victim could neither die nor stop the pain. Halyburton could take no more.

"*Bow cow! Bow cow!*" he yelled, indicating he was ready to talk to an English-speaking guard. Bug came in but was not ready to stop.

"You have not been punished enough," he said.

He tipped Halyburton over, stomped on his cuffs, and left. Hatred for Bug welled up inside Halyburton. His wrists were now numb except for nerves that ran along the side of each arm. The nerves felt as if hot lead were coursing through them. In fact, he swore he could see the molten lead flowing from his arms to his fingers, each drop creating more traumas. He feared that he would lose the use of his hands.

He was surrounded by anguish. He heard Navy Commander James Bell being tortured a few cells away, yelling for God to kill him. As Halyburton's own suffering intensified, he began to have the same feeling — he wanted to die. At one point, he wanted to knock himself out by banging his head against the cement floor, but he didn't have the strength or mobility.

Those reactions were typical among the Briarpatch victims. Both Navy Lieutenant Phil Butler and Navy Lieutenant Commander Robert Shumaker, finding their "hell cuffs" intolerable, tried to commit suicide by bashing their heads against the stone wall, while Ralph Gaither, who suffered lasting nerve damage in his hands, described his torture as "unspeakable agony of the soul."

Several hours after Halyburton had been knocked over, Bug returned and asked if he was ready to write, and the American said yes. His cuffs were taken off around dawn, but he couldn't move his hands until the afternoon. Even when he was ready to write, he wasn't ready to give in. He is left-handed, but he wrote with his right hand so someone reading it might conclude it was a forgery. He wasn't concerned about the biography; the Vietnamese already knew a good deal about him, probably from a news article. So he disclosed information they already had — that he had gone to Davidson College and that he was in VF 84 Squadron. The interrogators made him rewrite his draft several times, and Halyburton realized that what he wrote didn't matter — they just wanted to assert their authority. In fact, most POWs recognized that the

Vietnamese would accept virtually any story, however ridiculous, as legitimate. One Briarpatch captive, Air Force Captain Kile Berg, spoke so convincingly of Batman and Robin that his captors asked where the crimefighters lived and what political party they belonged to.

Halyburton was more concerned about his confession. He knew the Vietnamese were considering charging the Americans with war crimes, and he feared that a statement could be used against him. Halyburton rewrote his confession four or five times, believing he found a way to give his adversaries what they wanted while still protecting himself: "If my country has committed war crimes and this is an illegal war, then as part of the armed forces I am guilty as well." He believed that his use of the word "if" indicated that he himself did not accept that America had committed any such crimes.

While he may have won a rhetorical victory, he was still devastated by his capitulation. He knew he had done his best, and he knew that the U.S. military would not hold him accountable for his statements. But he believed he had let down the other prisoners, and so his resistance, however determined, had fallen short. This psychological blow was far greater than the physical injury. "It was about the worst time of my whole life," he said years later, "because I knew I had failed."

He felt somewhat better after tapping to Howie Dunn, who assured him that he was not a traitor and that he had tried his best. Moreover, he learned that most of the senior officers, including Jerry Denton, Robinson Risner, and James Stockdale, all men he deeply admired, had also given statements. That helped him begin to come to terms with his own sense of defeat. But for all the tortured prisoners who ultimately complied, the feeling of loss never completely faded. John McCain, the Navy lieutenant commander who became a U.S. senator, wrote of his own experience:

I couldn't rationalize away my confession. I was ashamed. I felt faithless, and couldn't control my despair. I shook, as if my disgrace were a fever . . . Many guys broke at one time or another. I doubt if anyone ever gets over it entirely. There is never enough time or distance between the past or the present to allow one to forget his shame. I am recovered now from that period of intense despair. But I can summon up its feeling in an instant whenever I let myself remember the day.

Halyburton was tortured twice more, though he withheld compliance for shorter periods of time. Once he gave a statement about his missions, the other time an apology to the Vietnamese people that began: "If I have caused any damage . . ." After two months his torture was over; the Vietnamese probably decided they had extracted as much as they could from a junior officer.

Ironically, the experience fortified Halyburton. His torture intensified his hostility toward the Vietnamese, his anger becoming a survival mechanism — what he called "an armor of hatred." As the years passed, his armor protected him from abuse, privation, and loneliness. Nothing could defeat his rage; survival was his ultimate revenge.

After ninety-three days in chains, Cherry was put back in a cell with Ron Byrne, who had fared no better. Locked in cuffs and irons and beaten daily, Byrne had lost thirty pounds, had welts on his face, and looked like "death warmed over." Actually, it was Cherry who was soon flirting with death. He began coughing and spitting up wet, sticky lumps. The cell was too dark to see what they were; at Byrne's suggestion, Cherry filled the white pot used for bathing and coughed into it. The water turned crimson.

"Damn, it's blood," Cherry said.

Byrne said he would call a camp official, but Cherry, having been beaten for the past three months, balked.

"If I ask them for something, they're going to ask me for something," he said. "I ain't going to give them anything."

Cherry didn't want to show any signs of weakness to anyone; despite his record, he felt he had to constantly prove himself to friend and foe alike. But Byrne, fearing for his safety, demanded to see the medic, who eventually arrived and saw the bloody water. Cherry was taken to a hospital, x-rayed, and returned to the prison, but heard nothing for three months. Then, in January 1968, an official came to the cell and delivered the startling news: a bone chip had floated into his lung and was close to his heart; he could kill himself with any sudden movements. He would have to undergo surgery to remove the chip, but until then he had to stay calm and restrict his head movements; he was even excused from bowing.

Cherry's shoulder had fused together by now, so he assumed the chip came from a rib during a beating. Having already experienced Hanoi's medical care, he feared another operation, particularly one so risky. Sedation scared him. He put his chances of survival at less than fifty-fifty, but he seemed to have no choice.

However, in a totally unexpected move, the camp's vice commander, named Lump for a tumorlike growth on his forehead, said he could receive an early release if he showed "a good attitude," declared that he had been treated humanely, and apologized for America's crimes against his country. The release would have been the first from the North; Cherry could have agreed to comply, watered down his propaganda statements, and gone home. Given his perilous condition, he could have said he accepted his release to save his own life, and doctors would have confirmed the botched surgeries and multiple infections and crushed shoulder and body cast and leg irons and the floating bone fragment that was literally a dagger pointed at his heart. Had he gone home, few would have marked him a coward or a turncoat.

169

"I'm not interested," Cherry told Lump.

"You could go home and see your family," Lump said.

"I will go home in turn," he said, meaning the order of shoot-down.

"But you are not well."

"If you want to release me solely on my bad health, without any agreements, then I'll talk to you."

"Maybe you make a big mistake."

On February 16, 1968, Hanoi released three prisoners — Air Force Captain Jon Black, Navy Ensign David Matheny, and Air Force Major Norris Overly — in a red-carpet sendoff to the peace activists Daniel Berrigan and Howard Zinn. Cherry was admitted to the hospital in Hanoi in late March. When he woke up, he learned that the doctor had sliced open his torso and cut out the seventh rib on his left side to remove the bone fragment from his lung. A plastic tube was protruding from his chest and draining fluid (blood and mucus) into a green beer bottle on the floor. The tube stayed in for days, and every time it filled a bottle, someone would dump it and bring back a new bottle. It wasn't painful, just bizarre; forever after, when he drank beer from a green bottle, he'd think about his days in that hospital.

One day in April a political cadre entered his room.

"Did you know Martin Luther King, Jr.?" he asked.

"Not personally."

"Did you know he's been murdered by the American imperialists? The whites have killed him," he declared triumphantly, rattling off the details and weaving political conspiracies into the tragedy.

It was the most pain Cherry had felt in the hospital — perhaps in Vietnam. It even caused him to choke up, a display of emotion that he had always denied his enemies. He said nothing but felt as if the bone fragment that had just been removed had returned to pierce his heart.

The surgery was not entirely successful: Cherry left the hospital with a hole in his side for the tube. The bandage did not adhere to his body, and he had to use drinking water and a dirty cloth to try to keep the opening clean. He also entered a difficult new phase of his captivity: he was placed in solitary confinement in the Office for a staggering fifty-three weeks, the longest consecutive stretch in seven hundred total days of isolation. By now his interrogators recognized that physical abuse would not get Cherry to confess to crimes or criticize his country, so they made their most persistent effort to appeal to Cherry's identity as a black man.

Given the racial violence in America, which was followed closely by the Vietnamese, such an appeal seemed timely. After King's assassination, race riots broke out in more than a hundred cities, with at least 19 people dying and 3,000 arrested. Even that violence paled in comparison to that of the previous summer, when racial hostilities erupted in 128 cities, including Newark, where 26 people died, and Detroit, where 43 perished. In all, 164 disorders resulted in 83 deaths, 1,900 injuries, and property damage totaling hundreds of millions of dollars. With that evidence — and with the rise of groups like the Black Panthers, who preached a new self-consciousness among African Americans — the Vietnamese again urged Cherry to condemn his country.

Interrogations would last four or five hours a day, and sometimes two Vietnamese would play good cop–bad cop. The bad cop was Lump, who would threaten and berate the prisoner. The good cop was nicknamed "Stag," an acronym for Sharper than the Average Gook. He was well read, particularly in black literature, and liked discussing Ralph Ellison's *Invisible Man*, Richard Wright's *Black Boy*, or the play *Raisin in the Sun* by Lorraine Hansberry. He knew more about Stokely Carmichael and Malcolm X than Cherry did and enjoyed lecturing him about the struggle of Negroes against their white oppressors.

During one session Stag said, "Xu, we will change your base, your foundation." (The Vietnamese called Cherry "Xu," which meant "little brass coin," perhaps because of his color.)

Cherry said, "You're trying to brainwash me."

"Oh, no," he replied with unintentional irony. "Not brainwash. We have to change your thinking."

He talked about lynching in America, naming specific victims, and he showed news footage of recent urban riots. He often used the word "struggle": the "struggle" of the Vietnamese people was like the "struggle" of the Negroes, and they should be "struggling" together. "You shouldn't be fighting for imperialist America because they're against Negroes," he said. "You can't go to the schools you want to."*

Cherry believed he was sincere — Stag genuinely didn't understand why Negroes would support their country.

"Yes, we have problems," Cherry said, "but they're not for you to solve. We'll work them out."

"You're the one who doesn't understand. How can you fight small Vietnamese people like this, kill their women and children?"

"I'm a uniformed soldier and color has nothing to do with it." In time, Cherry just tuned him out, ignored his words, and sat comfortably on the stool.

Resistance also flourished at the Briarpatch, where Halyburton and the other Americans relied on tactics more subtle than confrontational to reclaim their pride and self-respect.

The Vietnamese, for example, demanded that Air Force Captain Bob Lilly read anti-American news items, which would be recorded and broadcast on the loudspeakers in camp. They had

---

*The Vietnamese were not without prejudice. According to one POW, a guard walked into his cell with a photograph of Wilt Chamberlain, saying, "He looks just like a monkey. Where does he ever find a woman to satisfy him?"

made similar demands of POWs in other camps, and some of them had used desperate measures to rebel. At the Hanoi Hilton, Risner pounded his larynx with his hand and then dissolved his soap in a cup of water and gargled, hoping the lye would damage his throat. (It didn't.) Lilly, at the Briarpatch, initially refused but complied after being tortured.

When Lilly's broadcast came into Halyburton's cell, he and his new roommate, Air Force Captain Paul Kari, couldn't believe it. Lilly quoted an American official in an authoritative voice, but when he quoted a Vietnamese official, he used exaggerated accents — French, hillbilly, or effeminate. He also mispronounced names: Ho Chi Minh was "Horseshit Minh," while his close associate Pham Van Dong was "Fan My Dong." Wilfred Burchett, the Communist writer who visited the prisoners, was "Wellfed Bullshit," and Prince Norodom Sihanouk of Cambodia was "Prince No Good Schnook."

Halyburton and Kari — indeed, all of the prisoners — laughed hysterically. Lilly's mangled words were inspiring — not because they were clever but because he got away with it. Other acts of insubordination had prompted swift retaliation, but in this case the Vietnamese did nothing. It was one victory, however small, that helped restore the Americans' pride.

The prisoners found more efficient ways of communicating. The tap code was safe and easy, and Halyburton was so proficient that he didn't hear taps so much as he heard "patterns," with words just appearing in his mind. But the code was limited to communication between two people. Then one day Halyburton heard the taps but not from a wall. They came from a prisoner outside who was hacking wood with a machete, but hacking in patterns that made words. Suddenly, everyone in the camp could receive the same message, which would allow senior officers to send orders, enable torture victims to warn of their abuse, or simply permit POWs to announce their presence. The key was getting outside the cell and then being resourceful and creative. So when the prisoners had to wash and hang their clothes, they would snap them dry, and snap them again and again until the snaps, like the taps, delivered a message. Later, Halyburton dryly noted, "We pretended to be very vain about our clothes." When the POWs had to tend a small garden, they used the broom to sweep out messages, such as: "FH . . . V MV 6 Last Nite in Truc." That meant: "Fuck Ho [a common phrase] Vietnamese moved six POWs last night in a truck." These messages had as much psychological as practical benefit, for they gave the Americans hope that there was still room for resistance.

Even maintaining personal hygiene was a way of rejecting the moral and physical squalor. Halyburton and Kari, for example, received two jugs of water each day — barely enough to live on. But they were determined to take a sponge bath, so they resisted drinking until moments before the new jug arrived. Then each man slowly dribbled the remaining water on a body part and scrubbed it with crumbs of soap. The rule was, one body part a day. Any more, and they'd have too little to drink. It took a few weeks, but both men completed their sponge baths.

By the end of the year, Bug had been transferred to Hoa Lo and the torture at Briarpatch had ceased. Conditions improved a bit; inmates were allowed to sing holiday carols, play chess and checkers, decorate a tree, eat a turkey dinner, and even tape songs. Air Force Sergeant Art Black did a drum solo of "Jingle Bells," while Air Force Lieutenant James Ray sang "Puff, the Magic Dragon" and Ralph Gaither, using a Czech guitar provided by the Vietnamese, strummed "The Wabash Cannonball."

The Vietnamese, fearing that the location of Briarpatch made it vulnerable to rescue attempts, shut it down in February 1967, and heroic tales of "the Briarpatch gang" would long set standards for suffering and endurance. Halyburton was grateful to leave but was hardly optimistic. He had now spent two Christmases in captivity, and he began to think that his release would not come for another year or maybe even two. Deprived of mail, he had neither heard from nor written to his wife or his mother. His last weeks on the *Independence,* when he pined for Marty in his letters after being away for five months, now seemed far away, but his longing for her remained the defining force in his life. He also tried to visualize Dabney — how long her hair was, what clothes she was wearing. But he never had a clear fix on his daughter.

By the beginning of 1969, the Vietnamese finally gave up on breaking Cherry's will. They weren't going to make him talk, but they weren't going to kill him either. He remained in isolation, but he was taken to the hospital to have the stitches in his back removed, and his open wound was treated. In April he moved in with Air Force Sergeant Arthur Cormier, a dark-haired medic from New Jersey who had already roomed with him briefly.

In his three and a half years in captivity, Cherry had neither written nor received any letters. He assumed his family had tried contacting him, but only the prisoners who wrote could receive mail. Cherry didn't write because he didn't want to ask for anything. A bigger concern was that the bombing of the North had stopped, and Cherry worried that the Vietnamese would have no incentive to release the POWs. In fact, without the flow of new prisoners to provide information, he couldn't be certain that the war was even going on or whether the POWs might just be forgotten. He realized how naïve his initial assessment had been — that the conflict would be over in months — but now he wondered if it could persist for five, ten, or even fifteen more years.

At least his health was better, but there were always surprises. He tried to exercise, primarily by running in place, but he began

175

to cough up blood and mucus. One morning when coughing, he felt something odd in his throat. He spat into some paper.

"Art, you won't believe this," he said.

"What have you got?"

"Look at this."

It was a piece of black fishing cord. Almost a year to the day after his surgery, he had coughed up a piece of stitching. He shook his head in disbelief. He had no idea what purpose it served, but it seemed that his body was trying to purge itself of this entire experience.

Halyburton's return to the Zoo from the Briarpatch was hardly an improvement. The summer of 1967 was miserably hot, with temperatures in the cells routinely over a hundred degrees at night; even the mosquito net became unbearable. Halyburton took his down, preferring insect bites to anything that might restrain the movement of air. He lived in different cells in the Pool Hall with varying roommates, who would use pieces of cardboard or wastebucket lids to fan themselves. But each cell seemed to be in the direct path of the sun. Some prisoners found they couldn't lie on their backs because their eye sockets filled up with sweat, which burned their eyes, so they would sleep on their stomachs.

Halyburton contracted a heat rash, first on his body and then on the soles of his feet and the palms of his hands. Pus formed in some spots, and his skin burned. Water made it worse. Even the guards had heat rashes. About the only ventilation in Halyburton's cells came from two three-inch-square holes in the ceiling. As his only source of relief, he would lie naked on the floor with his head next to the small crack beneath the door. There was not enough water to drink, let alone bathe. He and one of his roommates, Navy Lieutenant Dick Ratzlaff, tried to discuss "cool things," like snow skiing or the beach. They also talked about the

Bible and theology, but the heat was so oppressive, they had little energy to converse. Mostly, Halyburton prayed.

In August, looking out of his cell, he noticed a familiar face — Al Carpenter, the A-4 pilot from the *Independence,* who had been shot down the previous November. He had been on the ill-fated Alpha mission and was one of the pilots to report that no parachute was observed after Halyburton's plane was hit. Halyburton sent Carpenter a message.

"I'm here. I'm sorry to hear that you're here."

Carpenter returned with: "It's great to see you, but I reported you KIA."

Killed in action. The news panicked Halyburton, who had been certain the Navy knew he was alive. He had been told that if his photograph had been taken — and by now it had been taken on the day of his capture and during the Hanoi march — the CIA would know he was a prisoner. His concerns eased a bit after Ratzlaff and others predicted his correct status was probably known by now, but what about Marty? Had she already buried him? Had she already remarried? She had a baby to support and a life to live. Or what if, by the time she learned he was alive, she had already been involved with someone else?

The questions grieved Halyburton. Almost two years had passed since he'd been shot down, and he had not been allowed to write letters. Nor did he want to; he, like Cherry, did not want to ask anything from his enemies. So other prisoners, trying to inform the military as well as Halyburton's family, began sending cryptic messages in their letters. Al Carpenter wrote to his wife about paying the premiums for an insurance policy, identified as PH-617514 — which stood for Halyburton and his serial number. Dick Ratzlaff wrote letters to his wife that mentioned Halyburton's mother's telephone number as well as "Julius," Marty's nickname for him. Air Force Lieutenant Joseph Crecca, in a letter

home, used Halyburton's "personal authenticator number" —
3363 — when describing a radio model.

Halyburton had no idea whether any of these messages would
get through and, even if they did, whether they'd be too late. His
greatest motive to survive, what he thought and dreamed about
the most, was the love of his wife and his reunion with his family.
Now, all that could be gone.

12

Change in Status

For her first eighteen years, Marty Halyburton lived with her parents, the next four with college roommates, and the next year and a half with Porter. But after being told that her husband was dead, Marty was on her own, with a daughter to raise. Her mother had passed away, her father was estranged, and she had no siblings. She had no credit card, no job, and no employment history. She had no place to live. Porter's name was on her Volkswagen title, it was on their checking account, and it was woven into her own identity. As a wedding gift, Porter's mother had given her note cards, beautifully inscribed: "Mrs. Porter Alexander Halyburton."

Her new identity came gradually. She had to fill out and sign endless documents from federal agencies — the Veterans Administration, the Social Security Administration, the Navy — but the paperwork of death helped establish her independence. Finances were not a problem; she received veterans' benefits and social security funds, and a life insurance policy for Porter also paid out. But she could not get a credit card by herself — an uncle had to cosign — and, without a credit record, she could not get a home mortgage.

Practical considerations guided most of her decisions. For example, she was knitting a green sweater for Porter when he was on the *Independence,* and she had written that it would be done when he returned. After hearing of his death, she finished the garment and gave it to a friend whose husband was the same size. Where to live was her biggest decision. She decided to stay in the Atlanta area, where she'd been passing through and where she already had some family. She settled into a two-bedroom apartment in the suburb of Decatur and met other young families, the mothers convening each afternoon in the courtyard with their children. She received a new identification card from the military, indicating that she was the widow of a Navy officer. She established her own checking account and transferred the title of her car.

The one constant in her life, Dabney, was a source of joy but also a reminder of her loss. Often, the child would do something cute (smile at a stranger) or momentous (take her first step), and Marty would think, I have to tell Porter about that. Then she would remember she couldn't. Other times, she would think that she would just wake up and discover it was all a horrible dream.

But her sadness did not demoralize her. She made friends, went to dinner parties, and soon began dating a textile salesman. He was ten years older, still single, and was eager to make a good impression, taking Dabney to the park or zoo. He was ready to settle down; Marty said she needed more time but was pleased with the friendship and glad that Dabney had a male influence.

She didn't follow the war carefully but was aware of the POWs' mistreatment, which became an odd source of comfort: at least Porter had been killed instead of captured.

In February 1967, a man who said he was with "Navy records" called Marty, said he wanted to check on her, and asked whether there had been any changes in her life. Had she remarried? Did she have a new address? Marty thought the questions peculiar, but she answered them and thought nothing more of the call. Several days later, a man who identified himself as Laverne Miller, with the casualty branch of the Navy, called to inquire about Marty's well-being. After some small talk, he asked for directions to her apartment. Marty, assuming he was stationed at the naval air station in Marietta, Georgia, began giving directions.

"Oh, no," Miller said. "We've come in from Washington and we're driving from Hartsfield Airport. How long will it take to get to your apartment?"

"About forty-five minutes," she said.

When she hung up, she walked into her living room, where her best friend happened to be waiting.

"Dot, this man is coming here from Washington," Marty said.

"What do you think he wants?"

Marty paused, then spoke flatly. "They're coming to tell me that Porter's alive."

Her words scared her. Sixteen months had passed since she'd been told that Porter had died; she had never doubted the Navy's account and had no reason to question it now. She assumed the government could not make such a mistake. Yet, for some reason, she knew that everything had changed.

Dot Weesner had gone to Marty's apartment to drop off her son, for whom Marty was going to babysit, but she agreed to wait until the men from Washington arrived. When the doorbell rang, Dot took Dabney and her own son out the back door so Marty would not be interrupted. Marty opened the front door to see six men, all in civilian clothes, including Laverne Miller, who introduced himself.

"We've got something to talk to you about," he said. "Would you like to have a seat?"

"I know what you have to say," Marty responded. "You've come to tell me that my husband is alive."

Miller's jaw dropped. "I've been worrying about how I was going to tell you, and you've just made it easier on me."

The men went inside and sat in the living room. Miller offered a cigarette to Marty, who rarely smoked but accepted it. The conversation was similar to the one in which she had been told Porter had been killed: the Navy had few facts or details.

Miller himself never indicated that Porter was alive. In fact, he specifically said that he didn't know whether Porter was dead or alive. Rather, he said, there had been "a change in status."

Porter was now listed as a prisoner of war, based on an intelligence report. Miller could say nothing about the report except that it was months old, nor did he have any information about his condition. All he knew was that Porter had been captured after his plane had been shot down and that he had been, at one time, a prisoner of war.

Marty put out her cigarette and asked for another.

Miller took out a package, withdrew a stack of eight-by-ten photographs, and handed them to Marty. They had been taken, he explained, during the "Hanoi march" the previous July, in which POWs were paraded through the city. He asked Marty if she could spot Porter. Her hands trembling, she picked up the pictures and began looking through them. She saw the grainy images of the Americans with their heads down, their eyes averted, their faces grim. She moved through the photographs rapidly, and when she finished she went through them again. The exercise was wrenching; she was horrified at the prospect of Porter's being subjected to that abuse, yet she desperately wanted to see him. She went through the pictures over and over, smoked another cigarette, and asked questions, including one about Porter's pilot, Stan Olmstead. Miller said he was still listed as KIA. She tried to make herself believe this was happening and that maybe touching the photographs would help, as if the feel of emulsion would transmit some new reality; but her hands continued to shake and the images began to run together. She never saw Porter.

The men stayed for several hours; they were afraid to leave her alone, for they could tell by her eyes that she was in shock. Marty smoked a pack of cigarettes.

Miller said that he believed it was in her best interest not to share this information with anyone beyond her immediate family, that restraint would improve Porter's chances of returning safely. He also gave her the addresses and telephone numbers of several POW wives, including Sybil Stockdale, who was organizing a group that was trying to raise the awareness of the prisoners' plight. Before he left, Miller assured her that he would stay in touch.

When Dot Weesner returned, Marty's emotions finally overwhelmed her. She broke down in her friend's arms and sobbed uncontrollably, and she spent the rest of the day at Dot's house. It took days, indeed weeks, to sort through her feelings. When she believed Porter was dead, she had convinced herself that he had been spared a fate worse than death. But now it appeared he was suffering that very fate, terrible but unknown. Loneliness? Abuse? Starvation? She had no idea. Then again, she wasn't sure he was alive. If the Navy erred the first time in telling her he was dead, why wouldn't it blunder again in telling her that he had been captured? And what about Stan Olmstead? Why was he still dead? Why was Porter captured but not he? Was that right, let alone fair?

She wanted to believe that her husband was alive, but she concluded that this news was worse, far worse. As a widow, she knew what to expect and she could try to move on with her life. But as the wife of a POW everything was dark, uncertain. Compounding her problem was the restriction against telling others. She felt paralyzed. She couldn't say "good morning" or "how are you" without fearing that she would disclose her secret; and if she did, would that jeopardize Porter's life? She thought it might. She canceled plans to visit her aunt in Coconut Grove. She couldn't stand listening to small talk from neighbors, all of which seemed trivial.

She went to church late and left early to avoid conversation. The only people she could talk to were the gas station attendant and the grocery store clerk. She knew it was crazy. She could not spend her life shunning friends and acquaintances, but she couldn't deceive them by pretending that her life was normal. Unable to function, she called her aunt and then went with Dabney to Saint Simon's Island in Georgia, where Marty and Porter had lived for six months in 1964. Marty took walks past their old house to try to recreate the feeling of being a couple again, but it didn't work. Each day she agonized. The next two weeks were the toughest in her life.

She did tell her boyfriend the news, calling him at his country club and pulling him off the golf course.

"I've got news for you," Marty said. "Porter's alive." She wasn't sure if it was true but decided it was the best thing to say.

He took the rejection in stride. "I've lost a lot of girls," he said, "but never like this."

On her return from Saint Simon's, Marty received two telephone calls, the first from Laverne Miller. They had been talking on the phone regularly, and he had assumed a paternal role for Marty, who called him Mr. Miller. He told her that the Navy was going to release to the public Porter's "change in status." Two other Navy airmen who had been listed as KIA — Robert Doremus and Fred Franke, who were both shot down on August 24, 1965 — would also be identified in a public statement as POWs.

"It will be in Porter's best interest," Miller said. "You're probably going to be called by reporters."

"How do you mean you're going to 'release' it?" Marty asked.

"It's going to be very low key," he said.

"You mean like the little square they have in the bottom of the newspaper —'Five Americans killed in Vietnam' — something like that?"

"Yes, probably something like that."

Miller never said why the Navy was releasing the names, and he told Marty that when she was called by the press or anyone else, she should not speculate on how the government got the information.

Fifteen minutes later, she received a call from a friend in Davidson, who told her that Porter's mother had undergone surgery that morning. Cancer had been found in her abdomen, and she might not survive the night. Marty packed the car the following morning, scooped up Dabney, and headed for Porter's home town.

It was the latest in a series of medical crises for Porter's family. Perhaps it was coincidental, but Porter's absence seemed to draw the life out of his mother and grandparents. His grandmother, who had been ailing for years, had died the year after he was shot down, and his grandfather, now in his nineties, had been admitted to a nursing home. Katharine, in her late fifties and suffering from severe arthritis as well as cancer, was in and out of the hospital, where Marty and Dabney had visited her several times. Marty never relished the experience, as Katharine showed her little warmth, but at least she could spend time with her only grandchild.

Five hours after Marty left her house, she reached Presbyterian Hospital in Charlotte. Katharine had survived the night and was in stable condition.

In her room, Marty leaned forward and said, "Porter is alive."

Katharine nodded her head, a grin tracing her lips. "I never thought he was dead."

Marty then drove to Davidson to see Robert and Carolyn Bourdeaux, whose son, John, had been Porter's best friend. Robert Bourdeaux was a gruff, hard-drinking, gregarious character who still worked in the real estate business despite being crippled. He was blind, he had rheumatoid arthritis and diabetes, and he spent much of his time in bed, where he would listen to the news on the radio and television and would bang his cane to get someone's attention. But he loved Porter like a son and now embraced Marty and Dabney, who had often stayed with the Bourdeaux family when visiting Davidson.

When Marty walked through the door after visiting Katharine, Bourdeaux was in high dudgeon. "What the hell is going on here!" he screamed. "You've gotten five long-distance phone calls!"

Just then the phone rang — again — and Marty picked it up. A woman, identifying herself as a reporter in Washington, D.C., asked if she could verify that Porter was alive. If true, this was a huge story, she said, as Porter, Doremus, and Franke would be the first back-to-life cases since World War II, when twenty-three Americans thought to be killed in action were reclassified as prisoners.

Marty had been told that reporters might call, but she had no inkling that Porter's new status would be considered a major story. (She also didn't know how the reporter knew where to contact her, though Davidson was so small that most everyone in town knew where she was staying.) Suddenly, she had to describe Porter's condition, even though she had been wrestling with that very question for weeks. She didn't hesitate.

"Yes," she said firmly, "my husband is alive, and his status has been changed from killed in action to prisoner of war. I'm really hopeful to see him again."

Her words sent Bourdeaux into a fit. He lunged forward, grabbed the phone, and spat out a string of obscenities. "Goddammit! What the hell is going on, gal?"

"Robert, Porter's alive."

Bourdeaux hollered, and suddenly Marty felt that same excitement. "This is just what you wanted!" he yelled. Until then, she wasn't sure, and she hadn't accepted that it was possible, but the telephone kept ringing and the reporters kept asking her questions, and she kept saying, Yes! He's alive! And her words made Porter — his revival? his rebirth? his resurrection? — more real. One reporter told her that the White House was having a 4 P.M. press briefing on the new POWs, so that evening she watched *The Huntley-Brinkley Report* on NBC, which showed sketches of the three captives, including Porter. Seeing the image finally dispelled her disbelief and oscillation. If his sketch was beamed out across the entire country on national news, she knew it had to be true. The telephone kept ringing and friends and journalists kept asking her questions and she kept talking until the phone was finally pulled from its plug at two-thirty in the morning. It was, Marty later recalled, the most exciting night in her life.

When she returned to Decatur, she received a package; it was the green sweater that she had knitted for Porter and given away. "I think you're going to need this," her friend wrote.

While Katharine survived her surgery, the cancer still crippled her body. Usually medicated but often in pain, she spent her remaining months in Huntersville Hospital, a former tubercular center near Davidson that was, according to one resident, "the place people would go to die." Coincidentally, one of her nurses was a woman named Julia Johnson, who had worked as a maid for the family and had watched Porter grow up. Now, holding hands, they would reminisce about his youth — how he would fix the backyard feeders so the birds wouldn't go hungry, how he would be a leader for other boys in the neighborhood, and how he would obediently eat the snacks his grandfather made for him each day after school. Johnson recalled Porter as having "beautiful hands and a gorgeous smile," and talking about him made it seem as though he were there.

"Miz Halyburton," Johnson said, "what would you have done if Porter had been the opposite of what you wanted?"

"I never crossed that, so I don't know," she said.

Johnson soon realized that Katharine was trying to survive for just one reason: to see Porter.

"Julia," she said, "I know Porter's alive, and I'm going to be here to see my son."

"Miz Halyburton, how do you know?"

"I feel it in my heart. My son is somewhere on this earth. He is alive."

Her father died in January of 1968, at the age of ninety-four. Two months later, on March 4, Katharine died at the age of sixty. She wore her son's pilot wings to the end.

The news that Beulah Watts was waiting for finally came in 1969. In January, film footage of prisoners attending church at the Zoo's auditorium had been released to the United States. In addition to examining the tape, the government extracted photographs that were used to identify the men. In March, copies of the pictures were sent to Beulah as well as Shirley, and Beulah confirmed what the Air Force had suspected: Fred Cherry was in the picture, he was a POW, and he was alive. The film was shown on the news in Virginia, and Leolia could sing out for everyone to hear: "My baby's alive!"

She believed her prayers had been answered, though she could not slow her own decline. Diagnosed with cancer, she reconciled herself to never seeing her son again in this life. Her friends thought that, knowing Fred was alive, she could finally rest in peace. Leolia died in May of 1970, at eighty-two.

While Fred was now alive in the eyes of his friends and relatives, he was still dead to his wife. Her insistence on his demise sur-prised Calvin Hightower, an Air Force officer who had been the best man at their wedding. He called her after he saw the POW photograph in a magazine.

"I saw the picture of him — that's Fred!" he said.

"No, it's not," Shirley said.

"But it's a picture of him!"

"That's not Cherry because he's dead."

She didn't tell her children of their father's status — galling punishment for a man who had paid dearly for his life.

13

The Good Life

While Halyburton and Cherry led separate lives in Vietnam, their experiences overlapped in one critical respect. Both discovered that their lengthy incarceration had a transforming, even uplifting, effect. As the years passed, they were forced to shun conventional ideas of happiness and success, to reappraise the meaning of their own lives, and to create a world very different from the one they had known. Trapped, brutalized beyond despair, each man eventually survived by finding a higher plane of existence.

In the early years, Halyburton tried to escape the present by living in the past. He recalled every memory, cherished every achievement, confronted every mistake, and underwent what he called "the catharsis of regret." His "atonement" for his imperfect past was to think of everything he wanted to do in the future. In that process, he began identifying "categories of interest" — the aspirations of a young man who felt he had squandered opportunities in his carefree youth, a wish list of personal goals that would validate his imprisonment, a way to ensure that no minute of his freedom would be wasted.

Ultimately, he listed seventy-seven categories of interest, which he memorized in alphabetical order (from Art to Wife). They reflected his interests in liberal arts (music and writing, history and philosophy) while introducing needs that were practical (investing, insurance, typing), personal (family, friends), and spiritual (religion, the Savior).

The list was also a way to organize his life in his own mind. Each category had its own "folder," and he would imagine sorting through the folders in a giant file, pulling them out and reviewing everything he had "written" on that particular subject. He carried an imaginary notebook and pen, so if he got a new cellmate who was knowledgeable about, say, money, he would ask him for tips, write them in his notebook, and then file the pages in his Finance folder. He was determined, once freed, to make every hour of every day meaningful.

There was only one problem. When his reveries ended, he was still in prison. His fantasies were possible as long as he was certain that one day he'd be liberated. Indeed, what had made his incarceration bearable was his confidence in his imminent release, his belief that America would use overwhelming force to win the war. He had always established deadlines for the war's end: his birthday, Christmas, or the onset of summer. But this faith created sharp emotional swings, what he called "a never-ending sine wave" of expectation and disappointment. His captors always told him that he would be in prison for ten or twenty years.

When Richard Nixon was elected president on November 5, 1968, Halyburton believed that the Republican, an outspoken anti-Communist, would escalate the military force against the Vietnamese. Instead, Nixon wanted to reduce air operations, withdraw U.S. forces gradually, and transfer the fighting to the South Vietnamese. (He did begin the secret bombing of Cambodia on March 18, 1969, though the POWs knew nothing of those attacks.) The president believed this policy, known as "Vietnamization," would benefit the peace negotiations already under way in Paris while placating the war's increasingly strident domestic critics.

While "Vietnamization" did receive strong support in America, it demoralized many of the POWs, who believed the carrot-and-stick approach was doomed to fail against an implacable enemy. Halyburton learned about the withdrawal of troops from his interrogators, and he began to think the Vietnamese were correct. He might be held for twenty years or even longer, a prospect that forced him to develop a completely different attitude.

He no longer projected release dates but decided that he would go home when the time came and there was no honorable way to rush that moment. After four years, he concluded that imprisonment was not a temporary cessation of freedom but was, simply, life. Previously, he had escaped the present by excavating the past and imagining the future. Now he was going to embrace the present itself: he was going to find meaning to his existence that had no relationship to his freedom.

This new phase was made possible by group living arrangements, which had begun in April 1968. Until then, Halyburton had either lived by himself or with one cellmate, but as the number of POWs increased, he was placed in a cell at the Zoo with eight other Americans. They lived in room 2 in a building called the Annex, where the cells were comparatively large (seventeen by twenty feet). No longer did communication require tapping through walls, nor was it restricted to one partner. Discussions flowed on politics, history, philosophy, and dozens of other subjects while opportunities blossomed for education, entertainment, and exercise. The Americans had a chance to create a kind of embryonic society that was unlike anything Halyburton had ever experienced.

The prisoners, all college graduates and officers, organized themselves along military lines. On the basis of rank and time of service, the commanding officer was Navy Lieutenant (j.g.) Glenn Daigle. Halyburton was number two, the executive officer, which involved him more in deciding the resistance tactics of the cell. Even more prestigious was that he was the cell's first shootdown, earning him the nickname FOG, the fucking old guy. His perseverance was a badge of honor.

Everyone was given a different job: exercise, education, recreation, morale, operations, and administration. The last one was the historian, charged with remembering, or writing when possible, everything that occurred in the cell. Some jobs were more important than others, exercise being one of the most critical. Each morning the entire group did calisthenics. Not everyone enjoyed it, but the peer pressure was too much to resist. Competitions were organized for who could do the most pushups, jumping jacks, situps, or leg bends. The Navy pilot Irv Williams, a fitness enthusiast from Florida, did 1,250 consecutive pushups, which he then surpassed several years later when he did 2,250. He said he did them to keep his sanity. Halyburton had strong legs, so his specialty was leg bends, his primary competitor being a broad-shouldered Air Force lieutenant from Baltimore named Bernard Talley. Halyburton did 200 leg bends to Talley's 500. Halyburton did 1,000; Talley, 1,200. Halyburton 2,000; Talley 2,500. And back and forth. Halyburton finally did 5,000 and thought that might be enough to win. Then Talley did 7,000. Halyburton finally relented, not because he couldn't surpass 7,000, but because he couldn't do that many without interruption from a guard.

Halyburton was proud of his physical progress. He had never done more than twenty-five straight pushups before; now he could do a hundred. He also developed the strength to walk on his hands, another first.

Exercising their minds was equally important, with each man serving as both teacher and student. Courses were taught on history, gardening, music theory, math, and foreign languages — the basis for a trilingual dictionary (Spanish, French, and German) that Halyburton would later assemble. Math instruction led to the creation of a slide rule made of bamboo. Halyburton's own specialty was literature, favoring Shakespeare, T. S. Eliot, E. E. Cummings, and such classics as *Billy Budd* and *Moll Flanders*.

Entertainment tended to be a nighttime activity. Each man was required to dramatize a movie or book, with creative liberties encouraged. Entertainment, not interpretation, was all that mattered. The cellmates also took turns "cooking" meals and snacks, describing the ingredients, preparation, presentation, and taste, and then offering them with great panache. Halyburton was considered a very good cook, though he occasionally bluffed. He had never made chateaubriand, but he did in Vietnam.

If they could create food, they could certainly create music. Mike Christian, an Alabaman who hadn't worn shoes until he was thirteen years old, liked country music and could play the guitar, so he transformed a bamboo fan into a six-string instrument. It made no noise, but he fingered the strings as if he really were playing while humming the chords; he taught Halyburton how to pick as well. Fred Purrington, a jazz enthusiast, used two sticks to play the drums and formed a combo with Christian. Music had always been important to Halyburton, and he had spent many hours in prison singing to himself. Now he began recalling the words to every song he knew. When he forgot a line, he asked the others if they could supply it. Once he had all the lyrics to a song, he would commit them to memory. Then — and for the rest of his captivity — he would ask his cellmates to teach him new titles. He ultimately devised a jukebox in which he catalogued almost a hundred songs to a specific letter and number. When bored, he'd look over his jukebox, say "B12," and sing Patsy Cline's "Born to Lose."

Christmas always tested the emotional strength of the POWs,

194

so it was fitting that the Annex cellmates tried to revive the holiday. Halyburton secretly got a stick, draped a green towel over it, and decorated it with tin foil so his cellmates could wake up to a Christmas tree. It was placed on a small platform of clothes and blankets beneath the window, out of view of the guards. Also on the platform were red socks, sent in packages from home, hung as stockings. The men had a Christmas service, sang carols, and exchanged gifts. One person received a rosary made of tin foil; another used bread dough sprinkled with brick dust to make a red ashtray and red dice. Halyburton, using a bone he found in his soup, made a cross for a gift; he received a cigarette holder made of bamboo. The men also exchanged imaginary gifts that were redeemed after they were released. Halyburton gave Irv Williams a fishing reel, while he received from Mike Christian a triangular brass plate engraved with DREAMS, ACTION, REALITY, which reflected his philosophy of life: your dreams are followed by action, which creates reality. Not every gift was so philosophical. Halyburton received from Rod Knutson a gold-plated Zippo lighter inscribed FUCK COMMUNISM.

Halyburton devised the cell's single most ambitious idea. Earlier, he had created a chess board out of toilet paper. Other prisoners had created backgammon boards by laminating toilet paper with leftover rice and using pieces of pinched bread, dusted with red brick dust and ashes, as checkers. Now Halyburton wanted to play bridge, a game that his mother had taught him and he had played all his life. To create a deck of cards, he cut — with a stolen razor blade — the rough brown sheets of toilet paper into fifty-two squares and marked them with cigarette ashes and brick dust. He then had to teach all but one person how to play bridge — a complicated task under ideal circumstances. The rapport between partners, the bidding conventions, the play of the cards to win the most tricks, the ability to finesse an opponent — the game can take years to master. But under these conditions, the challenge was part of its appeal. It forced them to think, it absorbed many hours, and it involved all eight men — two games of four each.

When, in a random search, the guards confiscated the cards, it appeared the game was over. They were, however, determined not to allow the enemy to take away their game, but how do you play cards without the cards? Halyburton had another idea. If they could imagine food and music and movies, why couldn't they imagine a deck of cards? A new game was born: memory bridge.

It was a slow game, to be sure. One person was designated the dealer, and he dealt by whispering to four others the thirteen cards each one held. Those four were known as the memory banks, and they kept the hands for four different players. The players consulted with their memory banks to play their cards and to keep track of the tricks already played.

All the facets of memory bridge took at least twice as long as the conventional game, and disputes inevitably arose over who held which cards and which cards had been played. Instead of bringing the men together, memory bridge threatened to divide them, so it ended after a few hands. But it did represent a huge psychological victory, confirming the ingenuity of the Americans and proving they still had the freedom to use their minds as they chose.

Throughout the prison camps, the Americans became increasingly adept at converting their meager resources into instruments of survival. They stole string, nails, matches, toilet paper — and anything else that could be used for hygiene or communication. String and nails, for example, could make a mosquito net, while burnt match tips could etch notes. Wood and chicken bone were turned into toothpicks; blanket threads became dental floss; one prisoner with a sore gum put a pinch of tobacco between the gum and his lip and got "instant relief" from an aspirin-like effect. Stolen red peppers were used to plug ratholes — even the rodents couldn't eat through them — while pieces of metal, string, and a tin can produced a mousetrap.

Halyburton was one of the more resourceful prisoners. For example, he could sew. With their clothes and blankets always tattered, Halyburton wanted to repair them, but their captors wouldn't give him a needle. So he saved a pork bone and rubbed it against a brick to create a tip. He then found a nail and rubbed it to a fine point, which he used to bore a hole in the thick part of the bone to create the eye. He pulled threads out of a blanket to mend his own clothes as well as those of the other prisoners. He later found a copper wire, which made an even better needle (it was thinner than the pork bone). When he received red socks from Marty, he sewed them to a shirt and made a dickey, which kept him warm in the winter.

No one questioned Halyburton's toughness or military competence, but Purrington still thought his close friend was miscast in combat, for his creativity and artistry set him far apart from the other aviators. "I didn't know what the hell he was doing in the back seat of a Navy airplane," Purrington said.

Another close friend, Irv Williams, believed that Halyburton was haunted by the randomness of his own survival and perhaps even by guilt — that he should live while his more experienced pilot had died. "We talked about everything," Williams said, "and I always thought that in the back of his mind he didn't understand why he got out and Stan didn't."

Halyburton spent more than two years in different cells in the Annex, and while some of his cellmates changed, he was always in a group of eight or nine. He was in a community that could support him intellectually, emotionally, and spiritually, and he discovered that he could create a meaningful existence that did not depend on his freedom or on the conventions of success — money, academic degrees, career achievements, or the baubles of modern life. He could stay another ten or even twenty years without fear that his time would be wasted. He knew that the power of his mind was greater than the force of his captors; there was no place he couldn't go or imagine.

He and Purrington, for example, took long cruises on a forty-foot Concordia yacht. Purrington couldn't afford one in the real world, but he could in his fantasies. The yacht had a perfect name, taken from a song they heard on the radio, Nancy Wilson's "The Good Life." Her words rang true:

> O' the good life; to be free
> And enjoy the Unknown.

Aboard *The Good Life,* the men sailed to Buzzards Bay, off the coast of Cape Cod, and Halyburton described luxurious ports of call and the raptures of the sea. Beneath the high sky, he would say, the winds pull us through pristine Hadley Harbor, nestled between Nonamesset and Naushon islands. There sailors are forbidden and few houses have been built; sheep, coyotes, and ospreys rule, and boaters sit on their decks and watch the deer roam, unafraid, on the shoreline. We lie out in the sun and grill juicy steaks, drink wonderful merlot, smoke long cigars, and listen to Nancy Wilson and Johnny Mathis, anything romantic.

Naturally, they need female companionship, so Halyburton decrees that Purrington's date is Dominique Francon (a name from Ayn Rand's *Fountainhead*), a beautiful and exotic woman with a rich cascade of black hair. Fred and Dominique are lying on the deck in the early evening, snuggling and relaxed, and Fred is touching the waves of her hair and the curves of her body, and the cool wind is picking up; Halyburton is establishing a perfect mood and building to a slow but inexorable conclusion when he says, "Hair is spilling over your chest."

Purrington, in the prison cell, looked up and said, "How did I get hair on my chest?"

"No! No!" Halyburton yelled. "Not your hair, stupid! Her hair!"

The question destroyed the mood and ruined the moment, though it was a venial sin. Halyburton just had to start over again, for there was no reason to leave *The Good Life*. It was certainly good for Purrington, who characterized Halyburton's imagery and metaphors as "nearly poetic." Porter always had a date as well. Asked later if he sailed with his wife, he paused and said, "Not exclusively."

Fred Cherry found his equivalent of *The Good Life* by juxtaposing the unity he discovered in the prison with the racial tensions he had experienced at home — and, for that matter, with the hostilities that were driving apart white and black soldiers in South Vietnam.

In America's first truly integrated war, black officers commanded white troops, young Army officers like Colin Powell gained experience, and African American casualty rates (12.5 percent) ended at just under the proportion of draft-age black males. But the integrated troops were not always harmonious. Young draftees, white and black, brought with them the racial discord that was roiling America, discord that was exacerbated by the stress of jungle warfare. White soldiers displayed the Confederate flag in hooches, bars, military posts, and at USO shows. Graffiti such as "Niggers eat shit" and "I'd prefer a gook to a nigger" appeared in bars and latrines; in 1969 two white soldiers erected and burned a cross before a predominantly black barracks at Cam Ranh Bay.

African Americans responded with expressions of racial solidarity. They weaved bootlaces into "slave bracelets," displayed black power flags, used a black power salute, and performed a ritual handshake called "dapping." Some wrote on their helmets "We enjoy being black." Race riots occurred in two different stockades in 1968, the first at Da Nang Brig, the second at Long Binh Stockade. As the journalist Zalin Grant wrote in 1969, "Past favorable publicity about integration of the troops" in Vietnam "has shimmered and disappeared like paddy water under a tropic sun."

Race could have been a flashpoint in the prison camps. Sixteen black servicemen were held in captivity, most of the white officers had never lived so closely with African Americans, and the enemy tried to exploit racial differences. Disputes often arose among the prisoners, but they were never about race, in large part because of Fred Cherry. As the senior black POW, he set the tone for the other African American prisoners, and he used anything, including humor, to ensure unity between blacks and whites.

One day, for example, he tapped on the wall: "The Vietnamese are very democratic. They're treating us all like niggers." Another time, during a rough interrogation, he realized the questions were intended for another black Air Force pilot, Thomas Madison. When Cherry got back to his cell, he muttered to the others in his block, "They've got two niggers in here, and they get the wrong guy and they beat me up."

Humor certainly sustained them through the lowest moments. After a failed escape by two Americans in 1969, Cherry, as a senior officer at the Zoo, was selected for torture and solitary confinement. Afterward, he was placed in a cell next to Air Force Lieutenant John "Spike" Nasmyth's. The two men tapped messages daily, though they would have been punished if caught.

One time Nasmyth began the conversation by asking, "You got time for a joke?"

"Sure," Cherry tapped back.

"Did you hear the story about the armless, legless guy who rang the doorbell at a whorehouse?"

"No."

"The madam opened the door and said, 'What do you want?' He said, 'I want to come in and get laid.' She said, 'How the hell can you get laid? You have no arms and no legs.' He said, 'I rang the doorbell, didn't I?'"

Cherry laughed loudly. As he later wrote, "Here I was, damn near dead from torture and infection, and there's this guy in the next cell who has never even met me in person, risking his ass to tell me a joke in tap code. You just had to know what they would have done to Spike if they had caught him communicating with me at that particular time. It was at that exact second I realized how absurd the whole world was, and that I wasn't going to let it get me down."

Cherry always took pride in his professional appearance, even in prison, and he encouraged his cellmates to do the same. If their razors were dull, Cherry told them to keep the blade wet to improve the cut — they had no shaving cream — and then dry it immediately afterward to stave off rust. That would also minimize the chance for infection. Respect for religion was important as well. Many of the POWs said that God played an important role in their survival, but Cherry reinforced his own moral authority by disapproving of anyone using the Lord's name in vain. If someone said, "God damn," he would respond, "Do you have to say that?"

Cherry's behavior motivated others to do anything to help him. When he was living with Navy Commander Theodore Kopfman and Air Force Major John Stavast, his shoulder wound was oozing pus, and he spent some days drifting in and out of consciousness. His cellmates feared that the infection would be fatal. They had no medicine, but Kopfman recalled that his grandmother used to make her own soap with lye and would use it specifically to sterilize clothes. If it could be used for fabric, why not flesh? "We're going to stuff that wound with lye soap," Kopfman told Stavast. They did, and two weeks later the wound was dry and clean.

Kopfman had never had such a close association with a black

man, but he would carry Fred to the washroom and help him bathe, something he could have never imagined himself doing. "Bathing a body that is black was different," he said, "but after about a week, I never saw his color."

In September 1969, two events improved the conditions for all the POWs in the North.

First, two recently freed prisoners, Navy Lieutenant Robert Frishman and Seaman Douglas Hegdahl, held a press conference at Bethesda Naval Hospital and described their abuse in detail, making a mockery of Hanoi's claims of lenient treatment.

"I don't think," Frishman said, "solitary confinement, forced statements, living in a cage for three years, being put in straps, not being allowed to sleep or eat, removal of fingernails, being hung from a ceiling, having an infected arm which was almost lost, not receiving medical care, being dragged along the ground with a broken leg, or not allowing an exchange of mail to prisoners of war are humane."

The second event was the death of Ho Chi Minh. The new leadership of North Vietnam, faced with mounting criticism in the United States and around the world of its treatment of the prisoners, reversed course and markedly improved their conditions. The random torture all but ended, and the POWs received better medical care and were given more blankets, cigarettes, mail, and food (three meals a day instead of two, larger bowls of rice, eventually canned meat and fish, and tastier soup). Prisoners in solitary confinement suddenly received cellmates. The requirement to bow was dropped. Penalties for communicating were lessened. As a group, the Americans slowly gained weight, healed, regained color in their skin, and returned to something approaching physical normalcy.

A year later, another important incident altered the course of the prisoners' captivity.

To accommodate the overflow of Americans, the Vietnamese had opened a prison near the town of Son Tay, twenty miles northwest of Hanoi. It was one of the North's worst camps, with filthy cells, horrible food, and predatory rats. It was nicknamed "Camp Hope." On November 22, 1970, a U.S. strike force raided the Son Tay camp, lighted up the sky with bombs, landed helicopters, fought off defenders, and returned without losing a single man. Unfortunately, they also had no prisoners, for they had all been evacuated four months earlier.*

At the time of the raid, Halyburton was in a new prison at Dan Hoi, a barracks ten miles west of Hanoi. Comfortable by comparison, it was the only compound built specifically for the Americans, with freshly painted rooms, showers, courts for volleyball and badminton, and even facilities to make instant coffee. It was called Camp Faith. Halyburton described it as "a country club," and had he stayed there, the rest of his incarceration would have been relatively easy. But after the Son Tay raid, the Vietnamese feared additional rescue attempts and moved the POWs from the suburban camps at Dan Hoi and Cu Loc back to the more secure Hoa Lo Prison in Hanoi.

This turn of events left Halyburton with mixed emotions. Back in Hoa Lo, he found himself in a compound that held 340 American POWs; it was the first time all of the aviators had been together. Called Camp Unity, it was also crowded and tense, with forty or fifty to a cell. The prisoners walked shoulder to shoulder, sleeping mats overlapped, and the guards, concerned about another raid, patrolled with hand grenades.

But Halyburton was elated by the raid, despite its failure. It showed that his government was doing something to help. His

*According to Benjamin Schemmer in *The Raid*, during the Vietnam War the United States launched forty-five POW rescue missions, but only one succeeded. On July 10, 1969, an Army specialist was rescued from a Viet Cong camp but he had been shot just before his rescue and died two weeks later.

hopes that Nixon, once in office, would strike quickly and boldly had faded. For two years nothing had happened. The Son Tay attack, however poorly planned, lifted the morale of all the prisoners.

Also lifting Halyburton's morale were the packages from home, the first arriving on February 12, 1969. Marty had sent it two years earlier, but the authorities had inexplicably held it. It included red socks, a roll of Life Savers, and four pictures of Marty and Dabney. The package told him two things: Marty knew he was alive and she hadn't remarried. Subsequent packages brought him vitamin pills, pipe tobacco and cigarette papers, soap, a toothbrush, toothpaste, and small towels.

Equally important were Marty's letters, one of which arrived on the personal note cards of "Mrs. Porter Alexander Halyburton." (Most of her communications came on special telegrams, designated by Hanoi, which had only seven lines.) For several years Marty was told she could not send any letters, but she was finally allowed to write one in May 1969. Porter received it seventeen months later. "My dearest Port —" it began.

> Where do I begin except that this is the day I have prayed for and dreamed of. There is so much I want you to know, but you're the writer not I.
>
> My love for you is as strong as our happiest days together and it has never waned. These four years have changed both of us, but I have no doubt we have survived the most difficult test and that ahead of us will be an even more wonderful life together than we could have previously known.
>
> You know me, the optimist . . . I won't deny I never worried, never became impatient, or never discouraged but all in all I've held my head high and felt one day this would all be behind us. Then I have had our wonderful, sweet, beautiful Dabney. She is all of the above and more . . .

Hurry home my love. I thank God for your safety and with his help may we put aside our separation. I love you my very most sweetest with all of my heart.
Your Marts Forever

Marty tried to protect Porter, telling him in a 1970 letter that his mother had just bought a new piano — two years after her death. And in 1972 he received a letter from Dabney, now seven years old.

Dear Daddy,
I love you Tery is my new dog He is very sweet I had a birthday cookout at Stone Mountain I am taking good care of mommy Teri and Henrietta Love, Dabney

The letters were both uplifting and sobering. While Porter had found *The Good Life,* he knew his real life was passing by.

Fred Cherry's sister had been sending him packages for years, but they were delayed until 1970, and even then he received only a couple. He became acutely aware of what others were receiving, particularly when he was surrounded by other Americans at Camp Unity, which increased his feelings of deprivation. His senses sharpened by his incarceration, he could smell a new bar of Dial soap in another cellblock — he could almost taste it — or the burning of tobacco on the other side of camp. The other inmates, aware of his misfortune, shared their packages with him, leaving pieces of soap, candy, or cookies in the wash area. When he was isolated and forbidden such treats, the other prisoners would try to divert a guard's attention and roll items under the door.

Support came in many different ways. At Camp Unity, Cherry met Navy Lieutenant Giles Norrington, who had heard about Porter's and Fred's "legendary status" shortly after the airman was shot down in 1968. Norrington easily identified Cherry — in their building, there was only one black among twenty-six inmates — but was surprised that Cherry wasn't bigger physically. In fact, he was shorter than Norrington, though the abuse he had endured was obvious. His shoulder was twisted and withered, and a huge surgical scar across his torso looked as if someone had tried to cut his body in two.

Norrington saw Cherry go through good days and bad, the good ones coming on the days when the humidity did not exacerbate the pain in his shoulder. But the bad days were agonizing. Cherry was suffering from a pinched nerve and terrible back spasms, and to relieve the pain, Norrington would gently massage his back, taking great care to relax the muscles. At night he and other men — Bob Barnett or Dick Vogel or Bill Robinson — would lie next to Cherry and take turns rubbing his back so he might find comfort and fall asleep. "He may have been small physically," Norrington said, "but the man was a giant to me and to all who knew him well."

Over the weeks, Cherry never complained, but Norrington assumed he was suffering because he occasionally saw a tear fall from his cheek onto the rice mat. What he didn't know was that that tear had nothing to do with pain. Cherry cried, but they were tears of gratitude, of disbelief. That so many men — all strangers, all white — would stay up all hours of the night, slowly massaging his back and receiving nothing in return, reinforced his experience with Halyburton and the other POWs while defying so much of his own history as a black American. What he had often been denied — equality, respect, recognition — was now given to him in abundance.

Cherry cried, just as he had cried with Halyburton. But they were tears of revelation. He had found a more perfect America in a prison camp than he had ever found in America itself.

14

Divergent Paths at Home

By 1969, Marty may have known that Porter was alive, but she feared he would be forgotten. Some wives believed their husbands might be abandoned completely.

Their concerns arose in part from the peace movement, which seemed willing to sacrifice the POWs to end the war. That tack appeared more plausible as monthly demonstrations, starting on October 15, stunned the country by their size and intensity. A noontime rally in New York's financial district attracted 50,000, while the New Haven Green drew 30,000. Some 50,000 found their way to the Washington Monument, and 100,000 appeared at the Boston Common. Both politicians and peace activists called for variations on the same theme: "unilateral withdrawal," "immediate withdrawal," or "a timetable for withdrawal."

The antiwar movement did no favors for the POWs, whose release would not be hastened by the unconditional withdrawal of troops or diminished pressure on the North. But the protesters weren't the only concern for the wives — some were worried about their own government, particularly the policy of Vietnamization. Sybil Stockdale, who founded the National League of Families of American Prisoners and Missing in Southeast Asia, wrote to Nixon that deescalation could sacrifice the POWs "because there may be no specific end to the war."

Begun as a kind of wives' club and support group, the League turned into a highly mobilized lobbying organization. Its principal goal — to raise Americans' awareness of the POWs and urge humane treatment for them — was not welcomed by the Johnson administration, which tried to downplay any discussion of the prisoners. But the League organized a drive that sent two thousand telegrams to the White House on Nixon's first day in office, and it found allies in the new administration, whose Go Public campaign for the POWs coincided with the wives' objectives. The organization sent delegates to Paris during the peace talks in 1969 and to other capitals (Oslo, Geneva, Vientiane) where the governments might be helpful. Having operated from Stockdale's home town of Coronado, California, it moved its headquarters to Washington, D.C., in 1970.

Marty's involvement with the League began in 1967 and deepened with time. Still in her apartment in Decatur, Georgia, she initially wrote daily letters to U.S. senators or congressmen and, with other wives, visited Senator Edward Kennedy and other members of Congress who they feared wanted to end the war at any cost. Their message was simple: America must not leave its boys behind, and the country should never accept an end to the conflict without a full accounting of all POWs and MIAs. The pleas were highly personal and emotional, but the war was so divisive, most of the politicians wouldn't commit to anything; they didn't even want to address the POW issue.

Marty did not consider herself an activist. She had no public speaking experience or media training, and she was not one to ask questions or press an opinion. Her involvement was strictly personal. But as the wives drew attention to themselves and their cause, the media sought out poignant stories of resilience and hope, and Marty was a compelling figure. As Betty Olmstead, Stan's wife, said, "She was just this real cute, vivacious little blonde, and she was raising Dabney by herself."

She was profiled in newspapers and on television, with images of Dabney on a swing set or in a sandbox and with descriptions of her husband's memorial service. Marty was a portrait of sweetness, vulnerability, and loss. In a 1971 documentary, she sat demurely on a couch and described the odd life of a young woman who is neither single nor partnered: "I can't really fit into any group socially. I'm sort of the fifth wheel for bridge. I don't fit with married couples or singles . . . People who meet me are a bit embarrassed."

Marty said that Dabney, now six, kept her father in her thoughts. "She knows that the time in Vietnam is just the opposite of the time here, so when it begins to get dark at night, she says, 'Daddy is waking up now.' And my thought is, 'What does he have to wake up to?'"

The time difference was Marty's one consolation. Each morning when she got up, she told herself that Porter was preparing for sleep and would be free of misery. That thought — when you're sleeping, you can't be suffering — helped her get through her own day, every day.

Marty's responsibilities grew. By 1969, she was named the League's coordinator for southeastern states, which required her to keep state coordinators informed of POW developments. She was also on the League's board of directors and flew to Washington every three weeks for meetings, sometimes at the White House. Like Marty, most of the wives had never visited the White House before, and the seat of national power was impressive. Each time they were carefully checked through the outer and inner gates, a Marine was holding the door for them at the west wing.

In 1968 Marty received a call from an acquaintance's father, an attorney, who had seen her in the press and asked if she would speak at a function. She agreed; the engagement was six months away, and she didn't give it much thought. But soon he called again. The schedule had changed, and now he needed her to speak the following night.

Marty had never given a speech before and didn't know whom she'd be addressing. She wrote down some notes but, scrambling to find a babysitter, had little time to be nervous. She assumed it would be a small group.

Picked up and taken to a hotel in downtown Atlanta, she walked into a ballroom with six hundred people and saw a bank of microphones in front of the podium. She had missed the dinner — probably a good thing. She discovered she was at a meeting of the American Bar Association, whose clout and prestige put her on edge. As she waited in the wings, the man introducing her reminded the audience who was supposed to speak tonight — "Bobby Kennedy."

Bobby Kennedy! Marty couldn't believe it, and she trembled as she walked to the lectern and began to speak. Knowing that Vietnam had polarized the country, she carefully avoided any opinion about the war. She didn't talk much about Porter either, as she didn't want to elevate his status above that of the other prisoners. She spoke instead about how she learned of his death and reclassification.

"When he was dead, I had talked myself into believing that he was better off dead," she said. "But now I believe that Porter being alive is much more difficult. I believe it's a fate worse than death."

She said she had not received any letters from her husband, nor had the government of North Vietnam ever released its list of prisoners. "Regardless of your feelings about the war, these are Americans," she said. "All we ask is that North Vietnam adhere to the conditions of the Geneva Conventions, that they identify the prisoners they hold, and they protect them from abuse. That's all we ask."

Perhaps her most affecting comments were about her daughter. At the time, two-parent households were the overwhelming norm, particularly in middle-class suburban communities; divorce was stigmatized, and single motherhood was beyond the ken of many people. Marty knew that talking about Dabney drove home the POW ordeal. It was not about troop withdrawals in a remote jungle conflict but about one little girl "who has never seen her father." That was the tragedy.

She ended her ten-minute speech by telling her audience not to pity her. "The time that Porter's been away has gone slowly for me, but I can't even know how it must go for him," she said. "I've got Dabney and everything else I need, except him. He has nothing."

She was astonished to receive a standing ovation. She couldn't believe that a room of high-powered lawyers, all strangers and probably all men, had completely accepted her. She realized she could speak anywhere and it would be easy.

Within days she began receiving other requests, from rotary clubs, church groups, county fairs, and college students. Her speeches were short — no more than fifteen minutes — but emotional. Audiences were always sympathetic, with professional and well-educated groups asking good questions. But in rural areas, well-meaning but less sophisticated crowds didn't understand how Americans could be held overseas. "Why don't you just call him up?" one woman asked her.

Marty was one of the few POW wives in Atlanta, so anytime a development occurred, the media called her, and she became an increasingly familiar face and voice. Strangers approached her on the street and wished her well. She was asked to appear at candidate forums so that would-be officeholders, both Democrats and Republicans, could bask in her celebrity. Georgia's governor Jimmy Carter, elected in 1971, appointed her to the Veterans Affairs Commission, and she was invited to receptions at the state house. One time, she mentioned to Carter that the League was printing POW bumper stickers, and the governor said that every state-owned car would display one. Georgia then printed its own POW stickers, and Marty happened to be at the state house on the day Carter unveiled them at a press conference. When she overheard that the governor needed a car for a photo op, she volunteered hers, a battered Audi she had on loan from the dealer. It was, for Carter, a publicity coup. Who better to promote your POW sticker than a POW wife? He and Marty were photographed and filmed putting the sticker on her car. As the prisoners' saga played out over the next few years, the governor's office called Marty a number of times for the latest developments and for advice on how Georgia could promote the issue.

It didn't matter that Marty considered herself a conservative and that she didn't vote for Carter. She considered him a caring man who wasn't one of "the good ol' boys." When he was elected president in 1976, his press secretary, Jody Powell, offered her a job in the White House. She would have taken it, but by then Porter had returned.

Marty was also noticed by Ross Perot, the wealthy Texas entrepreneur and later presidential candidate who was a fierce advocate for the POWs during the Vietnam War. He understood marketing and publicity, and in 1969 he placed ads about the prisoners in 117 newspapers. He also understood the value of their wives in generating support. He sponsored a half-hour documentary on the war that included the wife of a missing pilot, and he enlisted four other MIA wives to meet with officials of Poland, Sweden, India, and the Soviet Union. (Perot was seeking assistance for shipping food to the prisoners.)

Not long after her debut at the ABA meeting, Perot's advance men called Marty and asked her to speak at a POW rally in Atlanta. She then agreed to address a crowd in Richmond, Virginia, and was flown there on a private plane. The attention was flattering, the travel glamorous, and Perot did everything on a large scale, with all the cameras, lights, sound systems, banners, and media. But Marty's speech left her emotionally drained, and the stage-managed spectacle overwhelmed and exhausted her. Perot was also demanding, and Marty didn't want to be drawn into his whirlwind. After Richmond, she told Perot's staff that she could not attend his rallies anymore. "There are plenty of wives who don't have kids," she said.

Fears that the prisoners would be forgotten never came to pass, and Marty had a small hand in the war's most successful publicity gambit. It began in 1969 when several California college students decided to call attention to the POW cause. A local television host was wearing a bracelet bought in Vietnam that he said reminded him of the suffering the war had brought. The students adapted that idea to a POW bracelet, brass or copper, each one inscribed with a prisoner's name and shootdown date. The students were part of an organization called VIVA (Voices for a Vital America), which enlisted many of the POW wives to sell the bracelet. It became both a fashion item among teenagers and an emblem of patriotism, and Marty's celebrity made her an ideal saleswoman. Governor Carter ordered "Porter Halyburton" bracelets for all Georgia's state police officers. She received so many orders that VIVA just kept sending her boxes of the jewelry, and she kept selling them for two dollars each, piling the dollar bills into a box. After about six months, she counted the money — it was ten thousand dollars in cash. The program ultimately distributed five million bracelets.

On May 10, 1970, Marty received her first letter from Porter, which finally confirmed for her that he was alive. She cried from his opening words:

Hi Marty, I pray this finds you, Dabney, Mother and all in good health. I treasure the pictures you sent. You both look great and I am so proud of you and love you so much. I was not injured on ejection and am in good health. Are you working, teaching or what and where? Hope you have kept a diary, pictures, tapes, etc. especially of Dabney. Please send more pictures and take care of Dabney. Much love to all, Porter

After that, the letters came sporadically, sometimes several in one week, but then an entire year passed with nothing. The delays made each letter more valuable; every line was studied intently. Marty was concerned that his handwriting had changed and wondered if his left arm had been hurt, but the individual letters were simply smaller, which allowed Porter to pack more words on his six lines. He usually reassured Marty that he was healthy while asking for more pictures and for all kinds of items, including "good hard toothbrushes," "Cuticura soap," and "food like p-nut butter, honey, chocolate, fruit jam."

In a letter to Dabney on September 28, 1970, he wrote,

I know from your pictures when you were only four, how strong & beautiful you must be by now. Please understand we must be apart for a while, but someday when I come home, I promise to show you so many wonderful things in the world . . . I am so proud of you. Four kisses for you. Love, Daddy.

Before Porter was shot down, Marty would have never given a speech to a room full of strangers, never attended a formal party without a male escort, and never walked into a governor's mansion or a senator's office or the Oval Office and felt comfortable in the presence of a powerful man. Now she had done all of that. She had developed self-confidence and independence, and she had never let anyone see her "cry her in her beer." Just as Porter had adjusted and grown in his life in captivity, so too had Marty. She discovered that some POW wives could not make the adjustment. In an extreme case, one woman would not let her children celebrate Christmas or their birthdays because she believed they should suffer like their father. When the other women found something to laugh about, she would say, "I wonder what our husbands are doing now."

But the year of 1972 tested Marty as never before. In January, Nixon announced that Secretary of State Henry Kissinger had been negotiating secretly with the North Vietnamese, briefly raising hopes that the impasse could end. But on March 30 North Vietnam launched an offensive across the demilitarized zone, and on April 15 Nixon authorized bombing near Hanoi and Haiphong. For the rest of the year, the peace talks moved in fits and starts. In the heat of a presidential election, the safe return of the prisoners was one of the few things the country could agree on, but even that issue became political. The wives themselves splintered over their support of Nixon or George McGovern, the Democratic nominee.

For her part, Marty feared that Porter would die in Vietnam if McGovern was elected, that the Democrat would sacrifice the POWs in his desire to leave the country in haste. She felt nothing but stress during the final weeks of the campaign, and for the first time felt "politicized" (though she didn't campaign). Nixon's landslide victory was comforting, and the talks between Kissinger and Le Duc Tho resumed on November 20. Marty was optimistic that the Vietnamese, knowing they had to deal with Nixon, would finally reach an agreement. Instead, the talks broke off abruptly in December, and the massive "Christmas bombing" ensued.

This marked the lowest point of Marty's POW experience, a feeling shared by many of the wives. They demanded and received a meeting at the White House, where they were met by Kissinger.

Marty attended but let others ask the questions, many of them tough and bitter. As usual, Kissinger had few answers, but the frustration was palpable. At one point, he slammed his fist on the table and stood up; Marty thought he almost had tears in his eyes. "I can only tell you," he said, "that we're doing everything we can and your husbands are our top priority."

Marty believed him, but more than seven years had passed, and Porter seemed no closer to freedom than on the day he was shot down. She canceled her social plans and tried to hide her emotions from Dabney. She wanted to cry in her beer.

Beulah Watts wanted to cry as well. She knew that her youngest brother was being betrayed — not by the Communists but by his own wife.

Concerns first arose in 1967 when she and Shirley attended a meeting of POW families. At the time, Shirley was receiving Fred's combat pay of $1,432 a month, but she asked if she could have her husband declared dead for insurance purposes, even though there was no indication from the Air Force that he had been killed.* Later that year Beulah called the Air Force, asking if she could be allowed to send her brother the one Christmas package that each POW was allowed to receive. She explained that Fred's wife did not care to send one.

In March 1968, Beulah made the first of many formal complaints to the Air Force, reporting that a man was living with Shirley. She repeated this allegation in November, but the service took no action, the policing of marital conduct not being part of its

---

*Shirley declined to be interviewed for this book. Unless otherwise noted, the information about her comes from one or more of her children or from court documents in a federal lawsuit against the Air Force, filed by Fred Cherry after his release, regarding money that the Air Force had given his wife. Shirley was not a defendant in the lawsuit.

mission. Shirley, who was raising four children, was also in touch with the Air Force. She had authorized the service to deposit $200 a month from Fred's paycheck into the Uniformed Services Savings Deposit Program, which paid ten percent interest. But in October she made her first request for "emergency funds" from the account. For the duration of his captivity, she would make twenty-three such appeals, averaging $720 per request. The Air Force complied each time. When an official once raised a question about the withdrawals, Shirley met the query with "hostility." Ultimately, she withdrew eighty percent, or $16,560, of the full deposit. She cited a variety of reasons for the "emergency" money, including vacations and "cash stolen."

At the time, Shirley's boyfriend was living with her and the children in Norfolk, but the money was apparently not being used to maintain the house. According to Fred Jr. and Cynthia, they often did not have a working telephone or a running toilet. Rats ran across the kitchen floor, dirt was everywhere, and Shirley herself was often absent. "The four of us were raising ourselves," Cynthia said. But when military representatives visited the house to discuss the requests for money, Shirley would clean up, and the children would quietly stand at attention. "She was trying to get more money, so she made us look like Beaver Cleaver," Cynthia said.

Shirley still maintained that the children's father was dead. One day in December 1969, Cynthia came home from school and saw baby equipment in her mother's room.

"Who had the baby?" she asked.

"Your mother did," the boyfriend said.

Cynthia, nine years old, was shocked and devastated. "That tore our family apart," she said. Fred Jr. hated the baby's father. The fourteen-year-old had recently seen photographs of his father in his mother's closet and believed he was alive. He asked his mother why she'd had a baby.

"I'm in love and we're getting married," she said.

"But you can't get married. You're married to Dad."

"Your father's dead."

"He's not dead. I saw his pictures in the closet."

Then Shirley reprimanded him for going where he didn't belong.

Beulah, who kept track of the children, notified the Air Force that Shirley was pregnant. She feared that a baby would divert resources from Fred's own children and further drain his savings. Her fears were borne out.

In late December, Shirley asked the Air Force for $500. "I just got out of the hospital from having a serious stomach operation," she wrote. "I had it done by a private doctor, since I was a little afraid, and wanted the same Doctor all the time." Shirley was entitled to free medical care at a military hospital, and the Air Force should have known that the "stomach operation" was a ruse. But it still gave Shirley her husband's money.

Beulah was outraged. In a letter to the Air Force on November 17, 1972, she wrote:

> I had a talk with Mrs. Cherry soon after she had the baby. I asked why she had not used some of the "birth control pills." She told me she didn't want to use that stuff, the baby was what she wanted. She also told me she intended to marry the man. Whether she married him or not I don't know but she never gave up Col. Cherry's pay . . . I don't know what the Air Force regulations are, but I don't think a woman should be entitled to her husband's pay if she is having children by another man.

Shirley used the money to help her boyfriend buy a house and a Cadillac, and she gave him some of Fred's belongings as well, including golf clubs, a brass table, and a stereo. "Everything our father owned was in his house," Fred Jr. said.

At the same time, Fred's children were getting into trouble with the law. Both boys joined gangs and were arrested for fighting; Donald was later arrested for armed robbery. "Most of the money my mother had," he later said, "was used to keep my butt out of jail." The kids were moved around to different schools, dropped out, or were suspended. About the only constant was their mother's contempt for Fred. "She tried to brainwash us that he was rotten to the core," Cynthia said. But even Shirley admitted there was one thing her husband could do: fly. "Mom always said, 'He could land a plane with a crate of eggs on the wing and not break one,'" Cynthia said.

In Hanoi, Fred despaired at the dearth of letters and packages from home. When his first letter and package from Beulah arrived in September 1970, five years had passed without any word from his family. He was the last of the early POWs to receive a letter.

Beulah's correspondence did not describe the family turmoil, but the absence of any information about his wife and children left Fred bewildered and heartsick. He began writing letters home in 1970, begging his wife and children to contact him. He wanted their love and support, but he also wanted basic items like food and toiletries. In a letter to Shirley and the children on November 20, 1970, Fred wrote:

> On the occasion of Xmas I wish all of you a very "merry xmas and a Happy new year." . . . I want very much to see and do all the things with [the kids] that I'd dream about. I have not received a letter or package from you and the children. I don't know why, but I would be very happy to receive letters & packages from you & the kids. Send good candy, salami, cheese, potted meat, soap, toothpaste and brush, pipe tobacco, and other things in small packages or cans. Also send instant coffee. I had a badly broken shoulder and I have spent 3 months in the hospital since I have been here. I

219

feel fine & in good spirits. Please take good care of mom. I want the kids to be good Boy Scouts like Daddy and study hard to get ready for College. I love and miss you. All my Love, Daddy

Fred received more letters from Beulah but still nothing from Shirley. He feared she had died. She had high blood pressure, and he thought the stress of being a single mother had been too great and she had suffered a heart attack. He convinced himself that he had caused his wife's death, and he would carry that guilt until he returned and learned the truth.

Fred's letters could reveal little about his treatment, as they were censored by the Vietnamese. But Beulah took them to a graphologist, who examined the handwriting and concluded that Fred's shoulder had "made him uncomfortable while writing" but that there had "been little, if any, deterioration in general health or mental stability." Those words provided some comfort, but as expectations rose in 1972 that the war would end soon, Beulah attended meetings with other POW family members at Langley Air Force Base. Taking notes on a yellow legal pad, she heard experts talk about "reverse culture shock," the need of "family [to] help him reestablish self-esteem," expected medical, emotional, psychological, and dental problems, as well as financial, legal, and career issues. Beulah heard that no U.S. military prisoner had ever been held captive for more than four years — Fred was already into year six — and no one knew when the POWs would be returning or what they would face at home. All Beulah knew was that her brother's return would be painful indeed.

15

Operation Homecoming

On May 13, 1972, a convoy of sixteen canvas trucks pulled out of Hoa Lo and headed north. They carried 210 POWs in handcuffs, including Halyburton, jammed together "like college kids in a phone booth," as well as supplies and equipment to set up a new camp. The Americans had no idea where they were going as they traveled all day and night. Halyburton had a concealed kit with personal items, which included a razor blade, and he sliced the canvas, poked his fingers through, and saw the rice paddies, trees, and huts as the fresh air provided some relief from the heat. The trip lasted two and a half days and they made only two stops; the food — salty fish — intensified their thirst. Halyburton rode beside several fifty-five-gallon drums of gas, the fumes hanging in the air while fuel sloshed out when the truck made a hairpin turn on a bumpy road. Guards smoked in the back of the truck, and Halyburton thought it would be a hell of a way to end it all — a cigarette butt in the gas fumes, blowing them all to pieces.

The trip ended near Cao Bang, a mountainous region only 9 miles from China's border, 105 miles northeast of Hanoi. The prisoners moved into stone and concrete maximum security buildings, surrounded by a brick wall, a karst ridge, and barbed wire. Without electricity, the rooms were cold and damp, and with the sun setting behind the mountains early, it was dark for fourteen hours a day. Light was supplied by kerosene lanterns. One inmate compared the space to an aboveground dungeon. When meat was served, it was impossible to tell what it was unless attached to, say, a chicken leg. The bread was old; the beans, rotten.

It was eerie. Sometimes a piercing sound filled the air, and guards wheeled around and searched the trees, filled with squealing, poisonous lizards. The guards also wore knee-high "snake boots" and carried sticks to beat the creatures away. This alarmed the Americans, who discovered a drainage hole under one of the beds and realized it was big enough for a snake; they stuffed it with socks. When Halyburton was bitten by a foot-long centipede, he smashed it with a sandal — and six or seven pieces crawled off in different directions.

Like all the prisons, it was given an appropriate name: the Dogpatch.

The Vietnamese never said why the POWs were moved there, but in time a sound guess was made. In April, U.S. forces had resumed bombing near Hanoi and Haiphong, and then, on May 8, the air raids intensified and Haiphong Harbor was mined. The Vietnamese probably feared that a rescue attempt or even a misguided bomb could eliminate their most important bargaining chip — the prisoners. So they took about half the captives to the Dogpatch.

In the northern locale, it got colder sooner. Other prisoners had received long underwear and other garments in packages from home, but Halyburton had no winter clothes. The guards gave all the POWs a third blanket, which helped, but Halyburton was still cold. Then one day a guard walked into his cell, yelled "Hat" (his Vietnamese nickname, a derivation), and threw a green sweater his way. He knew it had been knitted by Marty: it had no label and had a V neck, which he preferred; he remembered that Marty had been knitting him a sweater when he was on the *Independence.*

Releasing the garment and distributing the third blanket were signs that peace negotiations were progressing; the Vietnamese did not want to return emaciated or sickly Americans. But in October, after another impasse, Halyburton was taken in for interrogation and handed a copy of the proposed peace agreement.

"Why won't your government sign this?" the interrogator asked him.

"I have no idea," Halyburton said.

After seven years of interrogations and indoctrination and abuse, he was convinced of one thing: a peace treaty would never be signed unless the bombing intensified. He had felt that way for years, and he prayed that Nixon would not ease up now.

On a hazy December evening, Cherry was idling in Building 3 of Camp Unity when the air sirens began to wail and the antiaircraft guns sent flaming missiles across the sky. Jet engines roared above while falling bombs rocked the earth and plaster fell from the ceiling. "Máy bay Mỹ! Máy bay Mỹ!" came over the loudspeakers, which meant "American aircraft! American aircraft!" Cherry, standing on his bunk and looking up at the sky, thought North Vietnam had been hit by a nuclear bomb.

The POWs were jumping and slapping one another on the back. "They finally did it," someone yelled. "They nuked 'em!"

Actually, America was dropping mammoth bombs from its B-52s in what would be the war's most devastating display of air power. Bridges were destroyed, arsenals blown up, fires ignited. The "Christmas bombing" was launched on December 18 and lasted eleven days. Nixon wanted to end the negotiating stalemate by creating "the most massive shock effect in a psychological context," and the barrage was undeniably spectacular. When a missile

223

struck a B-52 at 30,000 feet, an orange-blue flash stretched across the sky, the large burning pieces falling to the ground, in the words of one prisoner, "like fire being poured out of a pitcher."

Some of the POWs were afraid, knowing that one errant bomb could kill them all. But most of them celebrated during each attack, yelling "Thank you!" and "God bless America!" Their confidence was bolstered by the panic of the guards, who scurried for shelter or blindly fired their rifles into the air. The morning after the first attack, Hanoi was silent, the usual wakeup music, horns, and traffic all missing. The interrogators and guards asked the prisoners what they needed and delivered morning coffee. Fred Cherry knew he was going home.

In the end, the B-52s dropped 15,000 tons of bombs, though Hoa Lo itself received little damage. Fifteen planes were shot down and thirty-four crew members captured. Thirty-two died. The POWs believed the attack compelled the Vietnamese, fearing further raids, to release them. One crew member recalled that, for years afterward, "if there were ever POWs around, we never had to buy a drink." The Vietnamese had a different view of the attack. They celebrated it as the "Dien Bien Phu of the skies," a victory that led the United States to withdraw its troops, which paved the way for the Communist triumph in 1975.

The Paris peace talks resumed on January 8, 1973; the governments of North and South Vietnam and the United States, weary of the stalemate and mounting casualties, began to resolve the remaining points. On January 18, the Dogpatch was shut down and the prisoners were returned to Camp Unity. When Halyburton entered his cell, a message was written on the door: "We'll be free in '73." In the waning days of the war, the Americans played volleyball, read magazines, socialized, and ate. The turnkeys and guards were still "arrogant," Halyburton wrote in his diary, "which was good. I wanted to leave NVN with the same taste in my mouth that I had for all these years."

On January 29 the prisoners, called to the courtyard, walked out in formation. Film crews and cameramen stood by. The camp commander, a professional bureaucrat nicknamed Dog, read the peace accord first in Vietnamese, then in English. He said a truce had been reached, the war had ended, and they'd be going home in two weeks. The Americans stood in silence. No cheers, no handshakes, not even a thumbs-up. Their silence was one more act of defiance. They would not allow their elation to be captured on film and turned into propaganda. For Halyburton, caution was also a factor. After so many years and so many false hopes, he would not allow his emotions to soar, only to be disappointed again.

Before Dog dismissed them, he told the prisoners to "show good attitudes" until their release. Robinson Risner, standing in front, did an aboutface and yelled, "Fourth Allied POW Wing, atten-hut!" Almost four hundred men snapped to attention, their rubber tire sandals coming together. The "squadron commanders," returning Risner's salute, barked, "Squadron, dismissed!" The men fell out and returned to their cells, then broke out in laughter and tears.

In the days that followed, the war's end could be seen and felt. With no fear of air raids, the lights on the weather and communication towers burned at night, and fireworks streaked the sky with ersatz bombs instead of real ones. The Vietnamese hoisted their flag at Hoa Lo to signify victory — except the flag was raised upside down and had to be fixed. Inside the prison, musical performers staged a farewell show, though most of the Americans refused to attend. The inmates received haircuts, new clothes, and a bag for a towel, soap, toothbrush, toothpaste, two packs of Dien

225

Bien cigarettes, and matches. The Americans were also fitted for shoes — their first since they were captured — by standing on pieces of toilet paper and allowing a Vietnamese to pencil lines around their feet. Some men hadn't seen zippers, buttons, or shoelaces for years and now played with their wardrobe "like a bunch of little kids in a toy store." For their final dinner, wine was served.

The prisoners from the North and the South were to be released over a six-week period, the order determined by date of shoot-down and medical needs. (The Vietnamese moved up to the first flight two POWs who had willingly denounced America and co-operated with the enemy.) On February 12 — which was, as the prisoners later noted, Lincoln's birthday — both Halyburton and Cherry filed out of Unity and into the courtyard, where the POWs, wearing dark trousers and gray zippered jackets, lined up two abreast in platoons of twenty. Some came on crutches; three arrived on stretchers.

As the day had approached, many of the Americans discussed what they would do when released. A popular line was, "A cold beer, a hot piece of ass, and a two-hour hop in a single-seat fighter, in that order." Several men described how they would exact revenge in ways that would repeat the brutality of their own experience. Halyburton realized that hatred had taken over their lives, and their freedom, so filled with bloodlust, would be spoiled.

Halyburton wanted to shed his own "armor of hatred" so that the Vietnamese would never impair him again. As he was led out through the gates of the Hanoi Hilton, he turned for one last look and said silently, "I forgive you." It was not, he later said, a Christian act but his final act as a prisoner to ensure his survival as a free man. He was overwhelmed with relief.

In the morning fog, two buses marked with camouflage pulled

out of Hoa Lo and puttered through the downtown traffic. There were no journalists or cameramen to record the moment, and few bystanders took notice. One American said he felt as if he was sightseeing on a tour bus while another remembered "being subdued in the solemnity of our thoughts, almost hypnotized."

The buses crossed a pontoon bridge and stopped at an administration building on the outskirts of Gia Lam airport. A Red Cross tent was pitched nearby. The building was like an old bus terminal, with everyone sitting on benches, and the Vietnamese passed out tea and pig fat sandwiches. The food disappointed Halyburton — "Couldn't they do better than that?" he later asked — but, famished, he ate a sandwich anyway. Beer was then served, offering some consolation.

Shortly before noon, the men returned to the buses and watched a U.S. Air Force C-141 Starlifter descend from the clouds and land on the runway. It was an "electrifying" moment, confirming their imminent freedom, and the juxtaposition of the great white and gray transport aircraft, with its high tail and swept wings, and the two small shuttles spoke volumes about the two countries. One of the C-141 pilots said the tinny buses "looked almost like toys."

A grassy enclosure surrounded by a wrought-iron fence served as the reception area, and officials from the United States and North Vietnam waited for the prisoners at a small table. Also milling about were representatives from South Vietnam, the Viet Cong, and an international advisory commission. As the POWs stepped off the buses, many saw a familiar face — Rabbit. Air Force Captain Charles Boyd recalled, "He was holding a muster sheet and was calling off the names of the men he had tortured for years. He did not look up." One by one, the men walked past the table and saluted a U.S. Air Force colonel, signifying repatriation, and a U.S. serviceman escorted each one seventy-five yards to a plane. One pulled from his jacket a white canvas with a message in blue letters: "God bless America & Nixon." Everett Alvarez, in captivity for almost nine years, stepped onto the plane and walked past a blond flight attendant, who took his arm. "I felt like I might oxidize into thin air," he later wrote.

Cherry, on the first flight, was welcomed onboard, checked for any immediate medical problems, and seated. The plane raised its ramp, taxied down the runway, and lifted into the sky. Like all the POWs, he restrained his emotions until the plane was airborne and the pilot announced "Feet wet," meaning "over water." Then the men hugged, screamed, laughed, and cried, toasting their survival with juice and coffee. When a "liaison officer" mentioned that the Miami Dolphins had won Super Bowl VII, someone asked, "What's the Super Bowl?"

They were headed to Clark Air Base in the Philippines, a three-hour flight; to ensure that no Americans were left behind, the officers began asking for the names of all known POWs. Navy Commander James Mulligan recited the name, rank, and date of shoot-down for 450 men. The officers also briefed them on the elaborate welcome that awaited them. Originally called "Egress Recap," it was now heralded as "Operation Homecoming."

For Cherry, the jubilation masked underlying worries. He was now forty-four years old, and he had been in prison for so long; decisions on such matters as sleeping, eating, and smoking had been made by others, and he wasn't sure how he would adjust to freedom. It seemed that he hadn't made an important decision on his own, beyond resisting the enemy, in ages. At the same time, the years of indoctrination and propaganda had shaken his confidence in America. He had heard so much about its troubles — the race riots, the assassinations, the antiwar demonstrations — he didn't know what kind of country he was returning to or whether he, a conservative career military man, would fit in.

Meanwhile, he had no idea if his wife was dead or alive or what had become of his family. He felt as though he was experiencing his own *High Noon,* a moment of reckoning that would change the course of his life.

Halyburton was on the second flight, and when he boarded, he hugged and kissed an Air Force nurse — and noticed that she was wearing Marty's perfume. He also had mixed feelings, his joy muted by the ambiguous outcome of the war. He knew that the United States hadn't won and doubted that the government of South Vietnam would hold once American troops were withdrawn. If the Communists prevailed in Vietnam, what was the purpose of the POWs' sacrifice? He certainly did not consider himself or any other prisoner a hero. After all, they had been captured, they had been pawns for the enemy, and they had failed in their mission. He puffed a Marlboro, jotted notes in his diary, and began writing a poem about freedom.

In the days leading up to her husband's release, Marty received so many telephone calls from Navy officials, friends, and family that she was forced to hire someone just to answer the phone. She finally had to install a second line. She didn't know what time Porter's plane would land at Clark Air Base, nor did she know which network would broadcast it, but she was prepared. An Atlanta station had given her two televisions, which she stacked on top of her own so she could watch all three networks at once.

On the release date, a dozen people crowded into Marty's small den and waited with her. Each network, it turned out, would broadcast the return live, as the event riveted much of the nation. But the vigil dragged late into the night, the festive atmosphere giving way to exhaustion, anticipation surrendering to impatience. It was one more painful wait for Marty. Porter's letters had been reassuring, but they had also been censored; and she wasn't sure what injuries he had suffered or whether he'd show any effects of his imprisonment.

When his plane landed around four-thirty A.M. EST, Marty recognized Porter immediately. He stepped off the plane confidently. He walked well. He looked healthy. He shook a hand firmly. Marty screamed and cried and hugged her friends. Porter was free. He was on the screen for only about ten or fifteen seconds, but it was enough to reassure her. The only mystery was his sideburns. They weren't fashionable in 1965, but the new shootdowns had probably told him they were now in style.

The phone soon rang — a local station. The producer knew Marty from her television appearances and asked if she wanted him to replay Porter's few seconds of fame right then, so she was able to watch her husband a second time that night. In the coming days, she went to the station, sat in a room, and watched Porter's tape over and over again. She still needed to convince herself that he was free, and she never tired of looking at the step-down, the walk, and the handshake.

The following morning Marty told Dabney, by then in second grade, that Daddy would be calling soon.

"You don't have to go to school today if you don't want to," she said.

Dabney went to the refrigerator, where the lunch menu was posted. "No," she said. "I think I'll go to school. They're having hot dogs for lunch."

Cynthia Cherry was sitting in her seventh-grade class when a friend, carrying a newspaper, looked at her.

"Your dad is in the newspaper," she said.

"What do you mean?" Cynthia asked.

"He's on the list of POWs and he's coming home."

"My dad is dead."

"No he's not. He's in the paper."

"No he's not. He's dead."

The two girls argued until Cynthia looked for herself: "Fred V. Cherry." His name was in the newspaper, and she was giddy. She took the bus home, ran into the house, and told her mother the great news. "Guess what, Mom. Daddy's coming home!" She danced around, but her mother didn't say anything.

"She already knew," Cynthia later said. "Her expression was severe."

"The world she had built," Fred Jr. said, "was coming apart."

When the C-141 landed at Clark, Fred Cherry hoped to see an American flag. Stepping off the plane, he saw hundreds; the Americans were met by a cheering crowd of two thousand people. The men walked off the plane in the order in which they had been shot down. They waved and saluted as the well-wishers screamed, "Welcome home! We love you!" The senior ranking officer, Jeremiah Denton, stepped to a microphone and made a brief statement: "We are honored to have had the opportunity to serve our country under difficult circumstances. We are profoundly grateful to our commander in chief and to our nation for this day. God bless America."

On the first day of Operation Homecoming, 116 prisoners were released. By April 1, 591 were returned. In all, with more than a hundred military and civilian personnel classified as having died in captivity and with nearly a hundred who returned previously by escape or early release, almost eight hundred Americans spent time as prisoners during the Vietnam War.

Clean sheets. Fresh soap. Hot water. Good food. The base hospital was nirvana indeed.

The men ate prodigiously; three-fourths of the returnees were able to eat a regular diet. Steak and ice cream were the most popular items, while green salads, rarely seen in captivity, were piled on. Banana splits were in such demand that the kitchen ran out of bananas. Also served in large quantities were eggs, sausage, chili, pizza, chitterlings, collard greens, strawberries, and peaches. One serviceman ate an entire loaf of bread, one slice at a time, two pats of butter on each slice.

Halyburton's favorite meal was breakfast, so he ate eggs, toast, and sausage, regardless of the time of day.

Cherry's first meal was also breakfast. He told the dietitian, "One platter of scrambled eggs. One platter of sausage patties. Laid out on two plates."

"But, sir," she said, "it's five o'clock in the afternoon."

"I don't care," he said.

She brought the plates.

In addition to food and reading material, many of the former prisoners requested vitamin E, which was supposed to improve sexual potency. (The men were advised that there was little scientific evidence to support that belief.)

Though slightly underweight, ashen, or pale, most of the men returned in generally good health, a reflection of their improved treatment in their later years of captivity. But the physical toll was still evident. About one-third of the men had suffered fractures, including many of the vertebrae broken during ejection. Others were infested with various intestinal parasites, such as hookworm. The dim cells had strained the vision of many men, necessitating eyeglasses. Chipped teeth were a common byproduct of the beatings, while the rope treatment left almost ten percent of the returnees with "peripheral neuropathy," nerve damage.

Also noteworthy was the absence of amputees, leading some to believe that the Vietnamese simply shot those POWs who were badly injured.

Less easy to document were the emotional scars, but the damage for some was clear immediately. Two former POWs committed suicide in their first four months of freedom, and mental health experts at the time predicted that every prisoner who had had a long period of captivity would suffer psychologically. "It is not possible for a man exposed to a severe degree of abuse, isolation, and deprivation not to develop depression born out of extreme rage repressed over a long period of time," said John E. Nardini, an Air Force psychiatrist who was himself a POW during World War II. "It is simply a question of when and how the depressive reaction will surface and manifest itself."

Halyburton, now thirty-one, emerged remarkably fit for his seven and a half years in prison, checking in with a few chipped teeth, some parasites, and a sore back. He weighed 159 pounds, 21 less than when he was captured, but he had gained about 20 pounds from his low point. At five-eleven, he was an inch shorter, apparently the result of vertebra damage from the ejection. He was also lucky. When he was on the *Independence* and a wisdom tooth had begun to hurt, a medic said he would be okay for a few months but would then need to have it removed. The tooth, however, never bothered him in prison — a blessing, given that the Vietnamese used pliers in dentistry. Halyburton had his tooth removed on his return to the States.

When he initially arrived at Clark, he saw the other prisoners calling their families, but he was told he had to speak to the chaplain first. He assumed that his grandparents, both of whom would be in their late nineties, had died, and he feared his mother may have passed away as well, for Marty had rarely mentioned her in her letters. But the chaplain was slow to arrive, and when he did appear, he was uncomfortable telling him the news. Instead of talking directly, he spoke elliptically, and he irritated Halyburton to the point where Porter almost felt sorry for him. When the chaplain finally said it — your mother and grandparents are deceased — Halyburton was determined not to display any emotion, as if to deny the chaplain any chance to comfort him. He just wanted to be alone. When he returned to his room, the news sunk in — the people who had raised him and loved him had died; his entire immediate family, except for Marty and Dabney, was gone. He'd never had a chance to say good-bye, and he didn't know if they had passed away believing he was dead or alive. In his room, he allowed himself to cry.

He finally called Marty.

"I love you and I miss you and I'm okay," he told her.

She responded in kind. He said he had just been told about his mother and grandparents.

"Yes, they didn't want me to tell you about that," she said. She soon asked her first question: "Are you going to fly again?"

Halyburton didn't want to continue as a navigation officer, forever in the back seat, but he didn't want to retrain as a pilot. In truth, he wasn't sure what he would do next, but he knew what to tell his wife.

"No," he said. "I've had enough of that."

As Marty later said, "He hadn't been away from being a husband so long that he didn't know what the right answer was."

Fred Cherry's return had already sent alarms through the Pentagon, which had a thick file about his family: the death of his mother, Shirley's living with another man and having a baby, the depletion of his savings account, the legal troubles of his oldest son. Air Force Colonel Clark Price, an old friend of Cherry's, took these problems to Air Force Major General Daniel "Chappie" James, a highly respected black officer who had developed close ties with POW families.

"We have a brother who's going to face some strong music when he gets back, and he doesn't know what's going on," Price said.

"Is Cherry violent?" James asked.

"Not when I knew him," Price said.

James dispatched Price to the Philippines on a special escort mission.

When Cherry arrived at the hospital, he weighed 132 pounds, which was close to his weight when he was shot down, but he had gained about 20 pounds in the last couple of months and about 50 pounds from his low point. He had had open sores or infections for six years, but they had all healed, as had his broken wrist. He hoped the surgeons would be able to restore mobility to his left arm; he could move it only about thirty-five degrees in front and seventy degrees in back. But the bones in his shoulder, now tightly fused, had previously been infected, and the surgeons concluded that opening up the shoulder would be risky. So Cherry learned how to use one hand for such tasks as buttoning a shirt and changing a light bulb. He also returned with blind spots in his left eye and hearing loss in his right ear, the result of exposure to jet engines and slaps across the head.

When Price visited him in a private room, he brought a folder with the litany of woes, and he started with the easiest.

"Your mother died on May 28, 1970," he said.

"I already knew that," Cherry said. He had spoken to Beulah, who had told him only about their mother.

Then Price described the betrayal of his wife, the problems of his children, and the raiding of his finances: during his years of imprisonment, he had earned $147,184 in pay and allowances, but he now had $4,720.98 to his name. He absorbed each piece of news as he had the threats and taunts from the Vietnamese, without emotion or anger. He just made the same comment.

"I can handle that."

"I can handle that."

"I can handle that."

Clark also told him that his sons were in the Army, not in college as he had hoped, but he wasn't upset. He was just glad they were not in jail.

Some returning POWs were devastated by the breakup of their marriages, but Cherry believed that, given what he had endured as a prisoner, he could survive any personal setbacks. He never said an unkind word about Shirley; fearing the worst about her health, he was just relieved that she was alive.

For that matter, Cherry refused to criticize the Vietnamese, even those who tortured him. He said they were "just doing their job." His friends concluded that he was incapable of hatred.

"I was frustrated by his lack of bitterness," Price said, "but I've never heard him say a bad thing about anyone." As if to humanize the enemy, Cherry gave Price some cigarettes from Vietnam.

"They taste just like Camels," Cherry said.

He did receive one piece of good news in the hospital. Two months earlier he had been promoted to colonel. That meant more prestige, control, and money, but they did not compensate for his biggest loss: his piloting days were over.

When Halyburton went to the PX to buy clothes, he got a sense of how much America had changed. Escorted by a young ensign, he saw a bizarre array of bellbottoms, floral shirts, shoes with brass buckles, white belts, orange hot pants, and miniskirts. He later called Marty and told her he had gone shopping. That evening on the national news, she watched a story about an unnamed former prisoner shopping at the PX and wearing a garish outfit. As the camera zoomed in, she felt faint. "Oh my God, it's Porter." He was wearing plaid bellbottoms with a red shirt — for the last time.

*   *   *

One night in the hospital, the emergency bell rang, and someone yelled, "Fred Cherry's dead! Oh my God, he's dead!"

The nurses sprinted down the hall, opened the door, and saw Cherry lying on his bed, motionless, his hands folded over his chest. They tried to revive him as other servicemen watched in apparent disbelief. Doctors hustled into the room while orderlies rolled in special equipment. No one seemed to notice that his bed was surrounded by four burning candles and vases stuffed with flowers. Everyone was thinking, how would it look if a legendary survivor of the Vietnam POW camps died in an American military hospital?

Cherry wasn't dead but he was out cold, the victim of a drunken stupor and a practical joke. Some other former prisoners smuggled Scotch into their rooms — drinking was forbidden — and they had a party. His body long denied alcohol, Cherry promptly passed out, so his friends used the flowers and candles to turn his room into a funeral parlor and then alerted the nurses to his demise. Cherry soon woke up, dazed, but able to confirm he was not dead. The hospital personnel were not amused, but Cherry found the prank quite funny and consistent with the raucous subculture he had long inhabited: they were fighter pilots being fighter pilots.

The halls were covered with Valentine's Day cards, and Halyburton read one that touched him deeply. "Dear Sir," it began, "I sure am glad you're all done. I said a prayer every night, and it finally came true. Welcome home sir. I would have gave my life to get you guys out of there. But I don't think my parents would like it. I think you'll like being home with your family. I'm a six grader. Gary."

One card of thousands, it was later immortalized in a book of letters about Vietnam.

*   *   *

Halyburton and Cherry had not spoken to each other since their last night together more than six years earlier. They had kept abreast of each other's well-being from other prisoners, but both had worried about how the other had endured the torture immediately after their separation.

When they saw each other at the hospital, they embraced, bringing tears to Cherry's eyes. He was amazed at how good Halyburton looked; physically, he appeared to have been unaffected by all his time in prison. Halyburton, on the other hand, was startled when Cherry lifted his shirt and revealed the new scars he had accumulated. He reminded Cherry how he used to count his old abrasions to pass the time, but now Cherry had several more, including one large one.

"What the hell was that from, Fred?"

Cherry told him about the lung operation to remove the bone fragment. Halyburton lifted his hand and ran it over the scars, the old ones and new, and was once again amazed by his friend's durability. He expressed regret that Fred had had to suffer so much. Cherry explained that the surgeon had left a stitch in him that he coughed out a year later, but he didn't complain about the mistreatment. Now he could laugh.

On his return, each former POW was admitted to a hospital for additional tests, and Cherry hoped to settle at Norfolk Naval Hospital in Portsmouth, Virginia, close to home. Given his family problems, however, he was sent to Andrews Air Force Base, outside Washington. "I guess many people were afraid I might have been crazy enough to do something violent," he later said.

His plane, which included a group of former POWs, landed at Andrews on the night of February 17, and the men were met by a crowd. As the senior officer in the group, Cherry was asked to give a statement. He had written down some remarks on the plane, but when he stepped up to the microphone, he saw his sons standing in their Army uniforms, as well as nieces, nephews, and Beulah. He was silent for almost a minute, and his lip trembled as he struggled to control his emotions. Finally he spoke: "We have been away for a long time . . . We accept this break in life as a necessity . . . We accept this break because we had a job to do . . . And we did that job to the best of our ability . . . We have come back to you with our honor, our dignity, and our pride . . . We were able to do that because we kept our faith in our God, our president, and our country."

Beulah, crying, hugged him. "Thank the Lord you're home. Thank the Lord you're home."

He saw that Fred Jr. looked just like him, and he was proud to see his sons in uniform: if he couldn't defend his country anymore, he was glad they could. He was disappointed they were in the Army instead of the Air Force, but he believed they would carry on the family name. As they were walking, he said, "I'm an officer. You privates walk on my left."

Cynthia had last seen her father when she was five years old, and in her mind he was "tall, dark, and handsome." Now she went to see him with her mother and sister, and when she walked into his hospital room, she was stunned by his size.

"What happened? Did he shrink in Vietnam?"

"No, that's his height," Shirley said.

"No, he shrunk!"

Fred was wearing a robe. At first he was smiling, but then he cried as Cynthia ran up to him and threw her arms around him.

There would be no reconciliation with Shirley. According to Fred, she did not want a divorce (which would end her financial support from the military), but the marriage was clearly over. The family's house was in his name, but when he visited it for the first time, most everything was unfamiliar. His own stereo, golf clubs, and silver coin collection, as well as most of the furnishings from their house in Japan, were all gone and never reclaimed. When he saw Shirley's boyfriend's bowling trophy on a shelf, he flung it against the wall, putting a hole in it. It was the first time his children had seen him lose his temper. "I thought, 'He's got a little spunk in him,'" Cynthia said.

The breakup of his marriage also divided the children; Fred Jr. and Cynthia embraced their father while Donald and Debbie were closer to their mother and adopted her hostility toward him. Not long after Fred's return, Debbie said to him, "I wish you had never come back. You ruined everything."

Fred suffered his pain silently.

The breach has never been repaired. While the war itself did not destroy the marriage, Cherry's absence did contribute to the family's dissolution. As Fred Jr. said, "We were all POWs."

On the *Independence*, Porter told Marty that he did not want family and friends at the dock when he returned because their reunion should be their special time together. Now, as he prepared to fly to the Naval Air Station in Jacksonville, Florida, he said the same thing: he wanted to meet her in the privacy of his hospital room. He had seen other returnees get mobbed at the airport, on national television, and he did not want to share such a precious moment.

Marty waited for Porter's plane in the control tower, where she saw the airport fill up with banners, a band, and crowds. The plane landed and a handful of returnees, including Porter, walked off. As the senior officer addressed the crowd, Marty ran down the tower's stairs, got into a car, and was driven to the base hospital. She reached Porter's room moments after President Nixon had called to congratulate her, a gesture he made, no doubt, to many of the wives who had supported him. A furniture store had equipped the returnees' rooms with television sets, living room furniture, a bed, and other appointments, while Marty had added flowers, telegrams, and photographs of Dabney.

Porter arrived. They were together again.

His health was so good that he did not have to sleep at the hospital. He and Marty were allowed to stay at an apartment on the base, and, using a donated car, she could drive him around town, his license having long expired. He gave her a diamond and sapphire ring that he had purchased at the PX in the Philippines. Dabney had come to Jacksonville as well, but she stayed with a Navy officer's family so her parents could have some time alone.

After a day, Porter wanted to see his daughter, so they drove to the house where she was staying. She was playing with other children outside, and Porter recognized her from her pictures. She had short blond hair and wore a blue dress with short sleeves, white socks, and black party shoes. "I thought she was the most beautiful child I had ever seen," Porter recalled. He hugged her, told her how much he loved her, and gave her a portable radio.

The family loaded the car and prepared to drive to a friend's beach house. Maybe it was because Marty had talked so much about Porter or maybe some children are just unfazed by such events, but the first encounter unfolded as if the family had never been apart. Once in the car, Dabney turned to Porter and asked, "Daddy, can I sit on your lap?"

Not everyone cheered the returnees; the most strident opponents of the war still found reason to fault them. The Reverend Philip Berrigan called the former prisoners "war criminals," while Jane Fonda, disbelieving claims of torture, said they were "hypocrites and liars." But such attacks carried little weight amid the testimonials of strength, stamina, and patriotism. For a war that had torn the country apart, had helped drive a president from office, and

241

had ended without the conquering of territory or the removal of a government, the safe return of the POWs represented a scrim of redemption. They appeared on television and radio shows and were honored by the president in what was described as "the most spectacular White House gala in history." Major League Baseball gave each man a lifetime pass to any game. Mayors gave them keys to their cities. Car dealerships gave them their latest models. Airlines gave them free passage. As a veteran journalist who had covered the Korean conflict said, "That war had heroes and a somewhat sympathetic press. The Vietnam War had neither until now." Or, as the *New Orleans Times-Picayune* wrote: "The nation begins again to feel itself whole."

Davidson celebrated Halyburton's homecoming on March 17, a windy St. Patrick's Day that saw Porter's old street blocked off. Picnic tables were set up, banners hung, and flags distributed. A keg of beer was rolled out, and the front porch of the house that Porter grew up in was turned into a speaker's platform. Porter, Marty, and Dabney arrived in a new Ford LTD. Hundreds of people started gathering at 11:30 A.M. — Governor James Holshouser arrived in a black limousine — and amid a band's patriotic tunes, the Halyburtons appeared on the street. Porter wore a red turtleneck and was, according to the *Charlotte Observer,* "in danger of being hugged to death by an army of smiling women." Marty, her blond hair tousled by the wind, smiled as she moved through the crowd, never more than an arm's length from her husband. Dabney was at her elbow, carrying a teddy bear.

Will Terry, who had delivered the "meditation" at Porter's memorial service, spoke first. He said he was the only person he knew of "who preached at a person's funeral and then welcomed him back." He presented a gift to Dabney and then said to Marty, "I'm sorry we don't have anything for you but Porter."

Mayor Tom Sadler proclaimed it "Davidson's finest day," while Governor Holshouser, an alumnus of Davidson, said, "If you think things have changed in the political world, wait until you see some of the coeds."

Charles Lloyd had been one of Porter's favorite professors, a colorful scholar who knew every line of English literature, had a bushy mustache that covered his mouth, and had a pup named Martin Luther.

"Porter," he said, "I never knew how much I loved you until I thought you were dead. You would have enjoyed your funeral, Porter. Thank God you missed it." He then led a chorus of "For He's a Jolly Good Fellow" and three "hip hip, hoorays!"

Porter, overwhelmed, spoke briefly. "I don't think I can tell you exactly how I feel right now," he said. "I feel like I've come home."

He visited the burial sites of his mother and grandparents and was told about his own burial marker. After his true status was known, the owner of the funeral home had dug it up, kept it in the garage, and waited for his return.

"What should I do with it?" the owner asked.

"I haven't thought about it very much," Porter said. "I'll let you know."

A month later, on April 14, Suffolk, Virginia, had its homecoming for Fred Cherry. Early in the morning, crowds lined up four deep on the downtown sidewalks, with an estimated seven thousand people gathered for a parade and ceremony. The event had the usual trappings of a hero's welcome — the speeches from dignitaries, the bands, the banners — but the racially mixed crowd was unusual. As one newspaper noted, the area was "the home of conservative Governor Mills E. Godwin, Jr., whose known anti-black attitude . . . is an extension of the feelings held by most whites in this South Side Virginia city." But during the parade Cherry sat in the back seat of a white Cadillac convertible, his dress blue uniform decorated with ribbons, and whites as well as blacks threw confetti on him, lunged to touch his hand, and blew kisses. The car was mobbed at one point by fans who wanted autographs and handshakes. According to one account, as his car passed the Saratoga Street intersection, "five white women from a nearby beauty salon stepped out in front of the crowd, raised their hands and clapped with total enthusiasm."

Beulah rode in the next car, wearing a mink stole and smiling proudly.

At Peanut Park, Cherry continued to shake every hand and kiss every cheek. The Reverend C. J. Word, from the East End Baptist Church, compared him to the prophet Daniel, delivered from the lion's den through the Lord's intervention. The mayor of Suffolk gave him a framed key to the city while schoolchildren presented him with posters they had drawn.

As Cherry had done his entire life, he unified the people around him. That night, at a dinner for him at the National Guard Armory, the Suffolk Community Male Choir sang "Born Free" and "The Battle Hymn of the Republic," and the state's attorney general, Andrew Miller, said, "If Colonel Cherry's ordeal is over, then so is our own."

Attending was Bill Robinson, an Air Force sergeant who spent more than seven years as a POW in North Vietnam. Robinson was one of the few enlisted men in captivity, and Cherry helped train him in the camp to be an officer. After their release, Robinson was commissioned as an Air Force officer, with Cherry handing him his bars.

That night Robinson returned the favor, presenting Cherry with his own portrait, to be hung in the armory, the first ever of an African American. Robinson, who is white, later said: "He spoke like me. He bled like me. He hurt like me. His attitude was not — 'I'm black, give me ten points.' If he had any approach, it was, 'I'm black, take ten points away so I won't pass you so fast.'"

At the dinner, Cherry looked trim and fit, his crippled shoulder hidden beneath his uniform. Before five hundred people, he spoke in a low, modulated voice, taut with emotion. He did not need notes.

"I am an American fighting man," he said, "and I wear this uniform to protect you and your way of life. I would have given my life if necessary, proudly and honorably. I was tortured severely. I was severely ill, but they never broke me. They didn't because I had faith in God, in my country — and in you. If necessary, I will do the same thing again because I want America to be what you want it to be. I will not stand by and see any country trample over the United States without offering my body."

He paused and took a deep breath. "I want to thank you for the most memorable day of my life, and I love you."

Epilogue

C herry did not want to sue the Air Force — the institution to which he had dedicated his entire adult life. But in the late 1970s, the lawyer who handled his divorce persuaded him to sue the service in federal court for failing to safeguard his assets. He was told that the legal action could help him recoup his money, set an important precedent, and even bring cash settlements to other military husbands.

The suit said that the Air Force should reimburse Cherry for the money it gave to his former wife in response to false claims and other unnecessary demands. The military had failed to protect Cherry's interests, the suit alleged, even though it had been told of Shirley's pregnancy and of the neglect of her children. Cherry asked for $122,098.13 of the $147,184 he had earned.

He was hoping for a quick settlement, but the Air Force rigorously defended itself, drawing out the case into an unseemly spectacle between a military service and one of its most loyal officers. Finally, in 1983 the U.S. Court of Appeals ruled the Air Force had been remiss in handling Cherry's money and that Shirley had not been entitled to all of her husband's pay. The three-judge panel agreed that the children "were seriously neglected and that this situation, like the pregnancy, would have been obvious upon casual investigation."

Nevertheless, the Court of Claims, which determined the reimbursement, ruled that the Air Force had to pay Cherry only $50,000. The rest of the money, it concluded, had been justifiably distributed. After the lawyer took her fee, Cherry received $35,000 and doubted that the outcome would benefit any other military personnel. "I felt my time had been wasted," he said.

The early shootdowns' years of captivity had coincided with revolutions in music, clothes, sex, hair styles, and morality. They had missed the hippie movement, the race riots, and hot pants; they were unfamiliar with the youth culture, the drug culture, and the counterculture; and they knew little about Neil Armstrong, Joe Namath, or Janis Joplin. Before he was captured, one POW recalled, Playboy models still wore clothes, and a "demonstrator" was a test car. The time warp would have been jarring for anyone, but it was particularly difficult for military men, who were trained to respect authority and to spurn those who would undermine it. Many believed that long hair meant hippies, which meant antiwar, which meant the enemy — associations that most discovered were wrong.

Cherry had to face another challenge because of changes in the country's racial climate. Many younger African Americans were more militant and did not accept his message of black progress through education, hard work, and compromise. When he described how well he had been treated by white POWs, an acquaintance said, "Keep that up, and you'll be an Uncle Tom."

Confrontational tactics, Cherry believed, were usually counterproductive. When young blacks rioted in Washington, D.C., he described them as "street niggers." But his biggest disappointment had nothing to do with politics or protests — he could no longer fly. If not for his shoulder, he would have gladly flown combat missions again. Instead, he worked on intelligence assignments for the Air Force until his retirement, as a colonel, in 1981.

Beulah died in 1993, and Fred would soon be the last survivor of eight siblings. He established a government consulting business and worked well into his seventies. He was lucky to remain relatively healthy. One day in the middle 1980s he was searching all over his house for a cigarette and, not finding one, prepared to drive out in the rain to get some. Then he thought, Why should he allow a cigarette to have more control over his life than the Vietnamese ever had? After forty years of smoking, he quit that very day.

Cherry bought a home in Silver Spring, Maryland, and stayed close to Fred Jr. and Cynthia, as well as his grandchildren, nieces, and nephews. He rarely discussed Vietnam unless asked, when he would tell his story. When he speaks to high schools and youth groups, he delivers the same speech he gave fifty years earlier, telling the youngsters that they can achieve whatever they want through hard work, education, and discipline. Asked why he didn't stand up to the bigots, he says he preferred to work within the system, believing his job was not to topple institutions by his protest but to reform them by his example. "I felt the best way to make society better was by being better," he says. "Long term, I think I've done more as a role model for the way I did it than if I had gotten into a fistfight over a hamburger." He also explains that the hardship in his life — the poverty of his youth, the racism of the military, the brutality of his imprisonment — made him a better, stronger person. "It made me what I am today, and I wouldn't trade a day of it," he says.

Cherry has received close to forty medals, including the Air Force Cross, the second highest medal given by the Air Force. In 1981 the service commissioned a portrait of Cherry, depicting him in three different poses: the ace fighter pilot with a scarf blowing around his neck, his hand holding his helmet; the POW in a black smock, his hands crossed, his face showing dignity and resilience; and the freed prisoner, accepting a light for his first cigarette (an image from a wire-service photograph from Clark Air Base). The oil painting, by Harrison Benton, now hangs in the Pentagon.

But no picture could capture his contributions to the military and the country as both a pioneer and a prisoner. After Vietnam, the armed forces adopted many reforms that would help make the military a model of integration on a large scale; but by then blacks had already destroyed the myth that they lacked the skill, courage, or patriotism to serve in combat. Perhaps no one had crushed that myth more emphatically than Fred Cherry.

Senator John McCain said in an interview, "It is people like Fred, because of their capacity to forgive and forget, that have led to a lot of progress in race relations in the military and in America. I don't know how he forgave that kind of treatment . . . I never would have. He's a bigger man than I."

It took time for Halyburton to adjust to freedom. For one thing, he couldn't sleep on a soft bed. He would start on a mattress but end up on the floor. He was easily frustrated by picky eaters, who would not finish the food on their plate. He did not easily forgive antiwar activists and was mortified when Dabney became a Bob Dylan fan. In the early years, he depended on Marty to make decisions for him. Even trivial or routine decisions, such as whether to read a book or watch a movie, were difficult. He was simply out of practice.

They initially stayed in the Atlanta area, where Porter taught naval ROTC at Georgia Tech University while earning a master's degree in journalism. But he had few professional options in the private sector. He wasn't a pilot, so he couldn't fly for a commercial airline. Writing interested him, but he wasn't keen about beginning on a small-town newspaper with a family to support.

He did what seemed most sensible: he stayed in the Navy. By the time he returned from Vietnam, he had already spent ten years in the service and saw no reason to leave before qualifying for retirement benefits. He attended the Naval War College, in Newport, Rhode Island, initially as a student, then as a professor of strategy. He retired from the Navy as a commander in 1984 but continued to teach as a civilian.

Porter and Marty did not wait long to expand their family. Their second daughter, Emily, was born in 1974, and the son they dreamed about having when Porter was on the *Independence* arrived in 1980 — John-Fletcher William, named after Marty's cousin who was killed in Vietnam and Porter's grandfather.

The war lived on. Supporters who had bought "Porter Halyburton" POW bracelets began sending them to him; he ultimately received more than two thousand and hung them from the ceiling like a chandelier. On each February 12 he wears the green sweater that Marty knit for him.

Soon after his release, Porter gave a speech about his captivity, saying, "I wouldn't do it again for a million dollars, but I wouldn't take a million dollars for having done it." He wanted to tell Marty about his imprisonment, as either a catharsis or a search for understanding, but she didn't want to hear it. They had both suffered enough, she believed, and she did not want to relive his pain. "I probably hurt his feelings a couple of times," she said.

Porter found solace in a copy of Viktor Frankl's *Man's Search for Meaning*, the Viennese psychiatrist's account of his survival at Auschwitz. His ordeal had been far worse than Porter's, yet Porter could identify with it, particularly his description of how the prisoners reacted differently to their captivity. Just as some in the concentration camp became "Capos" in serving the Nazis, some POWs accommodated the Vietnamese for special favors. Moreover, Frankl believed that suffering itself had meaning beyond survival. "If there is a meaning in life at all," he wrote, "then there must be meaning in suffering . . . Without suffering and death, human life cannot be complete. The way in which a man accepts his fate and the suffering it entails, the way in which he takes up his cross, gives him ample opportunity — even under the most difficult circumstances — to add a deeper meaning to his life."

Halyburton used Frankl's ideas to elucidate his POW experience: men could survive their suffering as long as they found meaning in it. Fred Cherry's suffering represented one more test in a lifetime of crucibles uniquely experienced by a black American, and his ability to meet each challenge, particularly in Vietnam, validated his worth. For Halyburton, his suffering allowed him to live fully a "Christian life," requiring selflessness and sacrifice while elevating the spiritual over the material. He later wrote his "life statement," which he began in prison but revised several times after he was freed.

> I wish, at the instant of my death, to be able to look back upon a full and fruitful Christian life, lived as an honest man who has constantly striven to improve himself and the world in which he lives, and to die forgiven by God, with a clear conscience, with the love and respect of my family and friends, and with the Peace of the Lord in my soul.

In the 1980s, Cherry contributed to an oral history of the Vietnam War, which drew attention to his friendship with Halyburton. That led, in 1986, to their joint appearance at Davidson College. The two men had seen each other many times since their release, but on that day they had a chance to address an audience.

Both men, sitting in adjacent chairs, talked about the importance of unity and communication to their survival as they recalled their time together affectingly.

"Our friendship grew because of our common bonds," Halyburton said. "We were both Americans, even though he thought I was a French spy, and we were in the armed forces. And we went through some terrible times together."

Halyburton was calm and straightforward and even tried to downplay the emotional ties of the relationship. "It was nothing we ever thought about at the time," he said. "We were just two Americans in a tough situation trying to help each other out. It was as simple as that. If it hadn't been me, it would have been somebody else."

If he had stopped there, the crowd would have believed there had been nothing exceptional about their experience. But he was not quite done. "I must also say that our friendship" — he reached over and touched Cherry's arm — "is one of the most special" — and he suddenly stopped. Memories seemed to flood into his mind, and tears welled up. He couldn't speak. He reached out to pour a glass of water and took a sip. Composed again, he said, "It was one of the most special things in my life, and I better stop now."

Cherry confirmed that their friendship was special, "because without Haly, I wouldn't be here. That's important to me, that's important to my family, and I know that's important to you. It's important to the young people I go and talk to, because without him, I wouldn't be able to do that.

"As Haly said, others in the camp would have done the same thing. But this *was* done. It was only he and I in the cell. Had he not done anything, no one would have ever known. He didn't do it because someone was watching over his shoulder. He did it because he was a decent human being, and I was another human being with him."

Afterward, Halyburton told Cherry that he was glad Fred had finally seen Davidson and visited some of the places and met some of the people he had talked about in prison. Cherry thanked Halyburton for making him part of his life.

Halyburton often saw Cherry on business trips to Washington and stayed at his house on several occasions.

In 1990 Cherry visited Halyburton's own home for the first time, a two-story former summer cottage in Bristol, Rhode Island. Halyburton took him out to the back yard to show him his favorite conversation piece: his gravestone. It sat beneath a trellis entwined by grapevines, surrounded by birch trees, Japanese maples, and magnolias, overlooking a vegetable and flower garden. A bench sat nearby, and Narragansett Bay lay in the distance. The marker was etched with Porter's name, naval aviator wings, and the words KILLED IN COMBAT OVER NORTH VIETNAM.

"That's eerie," Fred told him.

"Yeah," Porter agreed, "but it's nice to look at it from the top down."

Sources

The origins of this book lay in my interest in the military's desegregation. The armed services is the country's premier example of how a large institution successfully integrates, and I believe the seeds of that success were planted in Vietnam, America's first conflict with a truly integrated military.

I had never heard of Fred Cherry or Porter Halyburton until the fall of 2001, when I read Gail Buckley's *American Patriots,* a history of the participation of African Americans in U.S. wars. It had three pages on Cherry, including several paragraphs on his unlikely friendship with Halyburton. I wondered what had happened, why each man credited the other with saving his life, and how these kinds of experiences — multiplied hundreds of times across the armed services — changed the military's culture.

Buckley did not indicate where Cherry or Halyburton lived or, for that matter, whether either man was still alive; fortunately, the Internet makes it relatively easy to find out, and I soon spoke with, then visited, both men. I learned that each was proud of his conduct in Vietnam, and they believed their experience together reflected the military's highest ideals of sacrifice and honor. They wanted their story told and trusted me to tell it, asking for nothing in return; they read the manuscript for factual accuracy, but the conclusions and interpretations are my own.

Cherry and Halyburton spoke with me, in person or on the telephone, for dozens of hours, and they introduced me to their family, friends, and colleagues, civilian and military. I interviewed about 130 people, and I spent several days in their respective home towns, Suffolk. Virginia, and Davidson, North Carolina, to understand how their childhoods shaped their friendship. Both men are pack rats, and they shared with me their letters, journals, photographs, audiotapes, videotapes, news articles, speeches, military yearbooks, maps, bomb summaries, and — from Halyburton — a POW color timeline generated by a computer. They also authorized the release of their military records from the National Personnel Records Center in St. Louis.

Veterans keep in touch through well-organized networks, which helped me find people who knew Cherry and Halyburton. For example, to understand Cherry's experience as a cadet in the Air Force, Randy Presley, a fellow cadet in the class of 52-G, sent me a list of 135 names; I tried to contact each one. I received about 25 replies, and my respondents, recalling events from fifty years ago, all shared remarkably similar memories of Cherry's pride and determination. I was also given the names and phone numbers of Cherry's instructors and commanders. A notice in *Retired Officer Magazine,* seeking comments on either Cherry or Halyburton, generated additional correspondence.

Interviews were critical in fleshing out the POW experience. Retired Navy Captain Mike McGrath, who was captured in 1967, gave me a list of the former prisoners from Vietnam still alive. I tried to contact all 536 of them. Many responded, and while some had never met either Cherry or Halyburton, others had vivid memories, always positive, confirming the "legendary status" that the officers had achieved.

I interviewed Marty Halyburton many times but also benefited from a trove of audiotapes that she and Porter had sent each other while he was on the *Independence.* Listening to them was like eavesdropping on a private conversation. I also listened to a tape of Porter's mother updating her son on Davidson gossip. When Porter returned, he recorded Marty's description of the day she learned that he had been captured: the words have a haunting quality as she relives her moments of disbelief, excitement, and despair. Even more eerie is the tape of Will Terry's memorial service, which captures Dabney's soft cries in the background and makes Porter's death seem quite real. A cache of letters written by Porter, Marty, and Katharine bring the family further into focus. The library at Davidson College also has a large file on Halyburton and his family, which includes letters that faculty members wrote to American and foreign diplomats on Porter's behalf.

I spoke with all four of Cherry's children, though I met only with Fred Jr. and Cynthia, who were much more sympathetic to their father. I spoke to Shirley once, and she declined to answer most of my questions. As noted in the text, most of the information about her dealings with the Air Force and her former husband's money came from Fred's lawsuit against the service. Beulah's letters to and from the Air Force were also helpful.

The literature on the Vietnam War is vast, and I am fortunate to have made a dent in it. The single best volume, as both a historical overview and a compelling narrative, is *Vietnam*, by Stanley Karnow. Neil Sheehan, in *A Bright and Shining Lie*, brilliantly examines the war's tragic spiral through a charismatic but deeply flawed Army lieutenant colonel. The political culture that produced the war is dissected by David Halberstam in *The Best and the Brightest*. An excellent oral history, capturing a wide spectrum of voices from the United States and Vietnam, is Christian G. Appy's *Patriots*. Another oral history, *Bloods* by Wallace Terry, profiles black veterans of Vietnam, including Fred Cherry. *The Things They Carried* is Tim O'Brien's fictitious but classic account of his experiences as a soldier in Vietnam. William J. Duiker's biography, *Ho Chi Minh*, describes the dual forces of nationalism and communism that thwarted America's military. While not appeasing his critics, Henry Kissinger, in *Ending the Vietnam War*, conveys the difficulty of freeing the POWs while achieving other military and political objectives.

Several books helped me understand America's lengthy air campaign in Vietnam, specifically *Thud Ridge*, by Jack Broughton, *Clashes* and *The 11 Days of Christmas*, both by Marshall L. Michel III, and *F-105 Thunderchief* by Dennis R. Jenkins. *City at Sea*, by Yogi Kaufman, describes life on a modern aircraft carrier. Benjamin F. Schemmer, in *The Raid*, recounts the Son Tay rescue mission. *Stolen Valor*, by B. G. Burkett and Glenna Whitley, exposes the war's profiteers and fabricators. In *Vietnam, Now*, David Lamb explains why the Vietnamese embrace Americans three decades after the war.

The most comprehensive book about the POWs in Southeast Asia is *Honor Bound*, by Stuart I. Rochester and Frederick Kiley. As government historians, they had access to former prisoners' debriefings, which are otherwise sealed, and their book is a lengthy synthesis of the entire ordeal. Vernon E. Davis's *Long Road Home* is a companion book, examining the public policy issues surrounding the POWs. Both accounts were released more than twenty-five years after the end of the war. The first major study of the prisoners came in 1975 in John G. Hubbell's *P.O.W.*, which focused on the early years of captivity and tended to celebrate the Americans' resistance. Craig Howes's *Voices of the Vietnam POWs* offered a more critical look, highlighting examples of dissension among officers and collaboration with the enemy. *Survivors,* by Zalin Grant, is a fine oral history of one prison camp in South Vietnam, and *Bouncing Back* is Geoffrey Norman's sturdy profile of several POWs in the North.

The most familiar prison memoir is John McCain's *Faith of My Fathers,* and Robert Timberg's biography, *John McCain,* is also revealing. Jim and Sybil Stockdale jointly wrote *In Love and War,* about their respective experiences when Jim was in captivity. Other useful memoirs were Ralph Gaither's *With God in a P.O.W. Camp,* Jeremiah A. Denton's *When Hell Was in Session,* Mike McGrath's *Prisoner of War,* Gerald Coffee's *Beyond Survival,* George E. Day's *Return with Honor,* James A. Daly's *Black Prisoner of War,* Larry Guarino's *POW's Story,* Norman A. McDaniel's *Yet Another Voice,* Jay R. Jensen's *Six Years in Hell,* Everett Alvarez's *Chained Eagle,* John "Spike" Nasmyth's *2,355 Days,* and Larry Chesley's *Seven Years in Hanoi.* Richard Stratton's experience was recounted in Scott Blakey's *Prisoner of War.* An oral history of the POWs in the Korean War is presented in Lewis H. Carlson's *Remembered Prisoners of a Forgotten War.*

Several books examined the military's integration, including *The Air Force Integrates,* by Alan L. Gropman , Bernard C. Nalty's *Strength for the Fight,* Robert B. Edgerton's *Hidden Heroism,* and *All That We Can Be,* by Charles C. Moskos and John Sibley Butler. Accounts of distinguished black commanders, all pioneers, can be found in the autobiography of Lieutenant General Benjamin O. Davis, Jr., *American,* J. Alfred Phelps's biography of General Daniel James, Jr., *Chappie,* and General Colin Powell's autobiography, *American Journey.* African Americans' participation in Vietnam is studied in James E. Westheider's *Fighting on Two Fronts.*

For young adults, Walter Dean Myers wrote a biography of Cherry, *A Place Called Heartbreak.* A fine description of the Lower Tidewater can be found in *The Great Dismal,* by Bland Simpson, and in *Readings in Black & White,* edited by Jane H. Kobelski, which includes health and poverty data for Suffolk and a history of the community's race relations. The story of Davidson is told in two books by Mary D. Beaty, *A History of Davidson College* and *Davidson,* while a black barber, Ralph W. Johnson, wrote about the town in his memoir, *David Played the Harp.*

I found helpful articles on the POWs in the *New York Times, Washington Post, Washington Times, Providence Journal, Stars and Stripes, Time, Newsweek,* and *U.S. News and World Report.* Cherry's return was covered by the *Virginian-Pilot,* the *Richmond News Leader,* and the *Suffolk News Herald,* while Davidson's mix of small-town ethos and highbrow culture was covered by the *Mecklenburg Gazette.* Halyburton's experience in Vietnam produced stories in the *Charlotte Observer.* The single best article about Halyburton and Cherry appeared in the fall of 1989 in the *Davidson Journal,* a college publication.

Ludwig Spolyar, a former Air Force psychologist, gave me two papers he wrote on the prisoners in 1970 and 1973, which described the difficult adjustment to freedom that each man would face. Information about their physical condition after their release came primarily from *Medical Service Digest,* an Air Force publication. Details about Halyburton's cruise on the *Independence* came from the Naval Historical Center at the Department of the Navy.

Acknowledgments

My first acknowledgment goes to Fred and Porter, who fielded my endless calls and e-mails, accommodated my visits, and always delivered on my requests for more documents and details. They never complained, even when they had to slowly walk me through how a plane is launched from a carrier or a bomb is dropped from a jet. Marty was unfailingly gracious, and I hope she is pleased that her own contributions as a POW's wife have not been forgotten. I also appreciate that their children, Dabney, Emily, and John-Fletcher, shared their thoughts with me. Despite his family's breakup, I know that Fred wants the best for his children, and I am grateful that Fred Jr. and Cynthia candidly described a difficult history.

I am part of a select group of authors who claim Houghton Mifflin's Eamon Dolan as their editor. In this book, he pushed me to explore every angle of a white man and a black man trapped in a cell and helped me streamline a narrative that weaves together disparate prison camp and homefront stories. My agent, Todd Shuster, immediately recognized the power of the story and helped me structure a proposal that became the blueprint for the final product. I am always grateful to my manuscript editor, Luise Erdmann, who understands the craft of writing as well as anyone I know.

My thanks to Betty Dyess, who remarried after her husband, Stan Olmstead, was killed. She shared memories of Stan and Marty and gave me a photograph of Stan and Porter. I'll always be indebted to Marion

Godwin, who escorted me in Suffolk, introduced me to Fred's friends and relatives, and gave me photographs of Fred. I am grateful to Mike McGrath for giving me information on the former prisoners and for spreading the word about my book. I also appreciated meeting John McCain, who interrupted our interview to cast a vote on the Senate floor but returned to finish our conversation.

I owe special thanks to the hundreds of former prisoners who responded to my inquiries. Even those who didn't know Fred or Porter encouraged my efforts. I also got a sense of the lasting bond among these men. The first letters and e-mails arrived with "GBU" at the bottom. More notes came in, and the same thing appeared.

GBU.

GBU.

GBU.

The letters, of course, stood for God Bless You, which was tapped out as the universal sign off in the prison camps. In a way, I believe these men are still tapping to each other, still finding strength in unity. I never learned their code, but I won't forget their message of resilience, sacrifice, and survival.

While I can't thank everyone, the following people helped make this book possible: Dick Abel, Phil Alden, Everett Alvarez, Margaret Bourdeaux Arbuckle, James Bailey, Bob Barnett, Carl Battjes, Laura E. Bethea, Osborne Bethea, Jr., Marie Fisher Bjorneboe, Nancy Blackwell, Taylor and Irene Blackwell, John Bourdeaux, Eddie Bracken, Al Britt, Evelyn Brown, Ron Bryne, Arthur Burer, B. G. Burkett, Murray Bywater, Al Carpenter, Debbie Cherry, Donald Cherry, Jennifer Cherry, Carolyn Collins, Quincy Collins, Michael Cooper, Arthur Cormier, Joe Crecca, Glenn Daigle, Jeremiah Denton, Francis Drew, Hans H. Driessnack, John and Luann Fletcher, David Geffen, William Giduz, Joyce Ann Gnatt, Harry Gronewald, Thomas Halyburton, Roy Hart, Emmett E. Hatch, Jr., Roy Henry, Calvin Hightower, James Hiteshew, Julia Johnson, Mona Jones, Murphy Neal Jones, David Jordan, Ed Kenny, Frederick Kiley, Lawrence Kimbrough, Rodney Knutson, Theodore Kopfman, John Kuykendall, Jim Lamar, Lewis S. Lamoreaux, Bob Lilly, VanLear Logan, Tom Madison, Paul Mather, Norman McDaniel, Gerald E. McIlmoyle, Martha Fortner McInnis, Sam Morgan, Giles Norrington, Albert W. Owens, Leland Park, Wookie Payne, William Peoples, John Pitchford, Randy

Presley, Clark Price, Leo Profilet, Fred Purrington, James Raeford, Bruce Rankin, Donald Robinson, William Robinson, Brenda Ryan, Samuel Spencer, Ludwig Spolyar, Don Spoon, Erskine Sproul, John Stavast, Will Terry, Bill Thompson, Deborah R. Thompson, Jon Thorin, Konrad Trautman, Terry Uyeyama, Dick Vogel, Marvin and Raye Walls, Betty Wally, Henry Ward, Claude Watkins, Mildred Watkins, E. W. "Bill" Webster, David Wheat, and Irv Williams.

I'd like to thank my father, Ed Hirsch, whose concern for social justice has influenced all of my work. I'm grateful, as always, to my brother, Irl Hirsch, his wife, Ruth, and their daughter, Barbara; my sister, Lynn Friedman, her husband, Howard, and their sons, Sam and Max; and my mother-in-law, Aileen Phillips.

My wife, Sheryl, makes all things possible in my life, including this book. Her love and support can be appreciated but never repaid. And to our children, Amanda and Garrett, to whom this book is dedicated, I have only one thing to say:

GBU.

Index